UNFOLDING

A

POST-ROE

WORLD

Francis Etheredge

with a new foreword by Kiki Latimer

En Route Books and Media, LLC
St. Louis, MO

Make the time

En Route Books and Media, LLC
5705 Rhodes Avenue
St. Louis, MO 63109

Cover credit: Sebastian Mahfood
Copyright © 2022 Francis Etheredge

ISBN-13: 979-8-88870-014-3
Library of Congress Control Number: 2022951485

No part of this book may be reproduced, stored in a retrieval system, or transmitted in any form, or by any means, electronic, mechanical, photocopying, or otherwise, without the prior written permission of the authors.

CONTENTS

Introduction 1

Revisiting the Publisher's Blurb 11

A Biography and New Foreword by Kiki Latimer: Biography; the New Forward: *On the Reality of the Human Person: One in Body and Soul; On the Dignity of Women; To Choose Life; On the Reasonable Recognition of the Person; On the Coercive Choice; On the Objective Beginning of Personhood; On the Good of the Family* 13

Excerpt from an interview with Elizabeth Kirk on Promoting Adoption in America 29

Part I: Refocussing the Heart of this Book

What is in a Title? *Unfolding A Post-Roe World: Conception: An Unfolding Content; Rights Are Integral to Human Existence; The help We Need to Understand What Is not so "Familiar; Conception: An Unfolding Content; Conception Expresses Both Uniqueness and Relation-*

ship; Questions; Is My Body Mine?; This Book; Five Parts..33

Part II: The Teaching of Experience: Law; Analogies; and Gardening

Comprising Chapters One and Two....................... 67

Chapter One: An Imaginative use of Gardening and Plant Life: Part I: The Science of Plant Reproduction; Part II: Teaching through the Everyday Experience of Growing Vegetables: *Tomato Seeds, Plants and Conception; The Early, Hidden Development of Both Tomato Plant and Embryonic Child; Conception and Growing Potatoes; The Integrity of Human Being; Plant Loss and Human Sorrow*......................................81

Chapter Two: A Unitary Beginning of One or Many: *Two Views: Immediate or Delayed Animation; The Possibility of a Single Answer to When Did I begin? Embryology: What is One Organism? Reverting to What is Original; A Difference of Words: Embryo and Child?* 103

Table of Contents v

Part III: Literary Truths and the Literal Truth

Comprising Chapters Three and Four 119

Chapter Three: Passing Through the Past to the Present: *From the "literal" use of an Image to the Truth of Embryology; Stage One: Taking the Comparison with a Plant to be Literally True; Stage Two: Movement, Sensation and the Rearing of Young; Stage Three: Rational Ensoulment; A Concluding Reflection: Towards Understanding Human Ensoulment; The Greatest Natural Transformation: The Unfolding of Conception* 123

Chapter Four: Scripture and Theology: Word and Dogma: *The Word of God and Dogma; A Variety of Witnesses to Human Conception – Beginning with Eve; Job; David; the Martyred Mother of Her Seven Martyred Sons; Mary: The Dogma of the Immaculate Conception and Human Conception* 137

Part IV: What is Certain and What is Uncertain about Conception

Comprising Chapters Five, Six and Seven 155

Chapter Five: The Teaching of the Church and the Problem of Uncertainty: *Prologue: A Modern Moment; Introduction: Who is My Neighbour; The Problem of Uncertainty in both Church Teaching and the 14th Amendment* 163

Chapter Six: On the Interpretation of Texts: Particularly "Amendment 14": *Amendment 14; Mr. Justice Rehnquist, Dissenting; On the Question of the Rightful Protection of Women; and The Principle of Determining an Appropriate Level of Legal Action* 181

Chapter Seven: An Answer to the Uncertainty of What or Who Exists at Conception: *What is the Experience of Women in Pregnancy?; The Witness of Each One of Us; A Discussion on the Teachings of the Catholic Church and the "Opinion of the Court"; A Clarification as Regards the Teaching of St. Thomas Aquinas; The*

Contribution of Revelation and Dogma; A Variety of Bioethical Declarations; Gravitating to a Consensus 201

Part V: Unfolding a Post-Roe World

Comprising Chapters Eight, Nine and Ten 227

Chapter Eight: Roe v Wade: *The Ongoing Arguments of Benefit to us All: Justice Beyond a Change of Justices (i); Viability is for Life (ii); Choice, Burdens, and their Alleviation (iii); Bodily Integrity, Liberty, Equality and the Constitution (iv); Brain Death and Abortion (v); Abortion and the Advancement of Women (vi); True Justice is Irreversible (vii)* 231

Chapter Nine: A New Beginning: *The Opinion of the American Supreme Court; A Book by John Strege: Hannah: The First Child Adopted as a Frozen Embryo; IT-EST Webinars Entitled "A Post-Roe World" (I) and "Bioethics and Law: Understanding the Nexus: Truth and Meaning in Constitutional Jurisprudence" (II) and The ongoing challenge: Is There a Right to Life in the American Constitution?; A Possible Comparison* between the Legislative Variations of American States and a Precept of the European Court; Coercion: Personal and Social; An Ectopic Pregnancy: Fear, Truth and Opportunity; Pope Francis on Justice, the Bond of Fraternity and Peace 259

Chapter Ten: A Conclusion in Three Parts: Part I: Lest we Forget Mother, Child and Father; Part II: The Wider Implications for a Post-Roe World: *Three Loci: To Know our Identity; a Move Towards the Further Denial of Being Human; and Change;* and Part III: The Disruptive Power of the Word of God: *The Human Unpredictability of an Answer to Prayer: The Will of God* 301

A Very Distinguished Testimony: Saint Teresa of Calcutta... 325

Where To Get Help? "Seek and you will find" (Mt 7: 7): Help with Pregnancy or Post-Abortion Counselling—Dr. Pat Castle: '*78% of post-abortion mothers said if they had encountered ONE supportive person or encouraging message, they would have chosen life.*' And, therefore, he founded "Life Runners": they

wear '"REMEMBER The Unborn" jerseys as a public witness in over 3,300 cities'[1] 329

A Testimony from a Man to Men: *An Unexpected Joy: An Unprecedented Pain; Joy and Pain Express "Relationship"; "Indelible"; and the "Rise of Articles on St. Joseph and Fatherhood* 333

Further Reading: A Variety of Prior Work on Conception And Specific Documents of the Catholic Church in English 341

---

[1] ITEST "A Post-Roe World" Webinar publicity, now with the WCAT TV presentation on it: https://faithscience.org/post-roe/.

INTRODUCTION

"There is one question: How can I help? And there are many complementary answers. I hope, therefore, in one way or another, that both the question and the many kinds of answers are helpful."

As regards the crises of our times with respect to the problem of ever reaching an agreement about the nature of the human embryo, consider these statements by the *Supreme Court* of America and the *European Court of Human Rights* as opening an implicit dialogue, as it were, with this book:

The *Supreme Court* of America said:

'abortion is fundamentally different, as both Roe and Casey acknowledged, because it destroys what those decisions called "fetal life" and what the law now before us describes as an "unborn human being." [Footnote 13: Miss. Code Ann. §41–41–191(4)(b) (2018)][2].

The *European Court of Human Rights* said:

'that while the notion of property could not be limited merely to material objects, human embryos could equally not be reduced to the level of an object'[3].

---

[2] DOBBS, State Health Officer of the Mississippi Department of Health, et al. v. Jackson Women's Health Organization et al.: https://www.supremecourt.gov/opinions/21pdf/19-1392_6j37.pdf, "Opinion of the Court", p. 6.

[3] While I disagree with these authors at a variety of points in this book, there is a tremendous helpfulness to their scholarship. Anna Alichniewicz and Monika Michalowska, p. 131 of "Medicine of the Beginning of Life. Bioethical and Philosophical Arguments in the ART Debate", Warszawa, 2019: Appendix: Case Illustrations: European Court of Human Rights, Case of Parrillov. Italy, Judgment, August 27, 2015, http://hudoc.echr.coe.int/eng?i=001-157263 (accessed October 15, 2019) – as per the original author's document:

We live in a world, then, where principle and good practice are a witness to the gift of each person's life and their embodied right, expressed in the very fabric of human being, to integral human development and all that fosters their fruitful contributions to society.

But we also live in a world-wide community in which, to put it bluntly, there is the exploitation of the very substance of human beings with impunity, whether by stealing organs from the living[4] or bringing human life to exist in order to "mine it": the "it" being a girl or a boy or selling off an aborted child's parts. Who would believe, with the indignation at the offences committed during the *Second World War* that we would be living in an age where

---

https://www.academia.edu/42104311/Medicine_of_the_Beginning_of_Life_Bioethical_and_Philosophical_Arguments_in_the_ART_Debate?email_work_card=view-paper.

[4] Ralph Weimann, *Bioethical Challenges* at the *End of Life*, Brooklyn, NY: Angelico Press, 2022, p. 203, but this issue is raised also with respect to a controversial definition known as 'brain death'. Cf. also Francis Etheredge, *Reaching for the Resurrection: A Pastoral Bioethics*: https://enroutebooksandmedia.com/reachingfortheresurrection/.

countries. corporations and, ultimately, individuals can act with a disregard to the basic question of the identity, dignity and rights of the human being?

On the one hand, there is so much good science which testifies to the humanity of the scientist, the beneficial nature of discoveries and the growing wisdom concerning the first instant of fertilization as the first instant of conception[5]. On the other hand, this witness is like a candle in the night, flickering because of the prevailing breeze and yet it is a light and expressing the *Great Light* of the life, death and resurrection of Jesus Christ. Thus, the light that illuminates the gift of life expresses an illumination that enables us all to live and hope and love.

It was argued in the first edition of this book that the Supreme Court's task, to identify the original intention of the American Constitution, has wide implications for the plight of the preborn, world-wide, precisely because it brings to light a whole range of issues which need to be addressed by a just law; and, at the same time, it raises the question of the value of life and, therefore, the value of a legal recognition of the beginning of human life. As the

---

[5] See the end of this book for a variety of work on this subject.

Supreme Court's Opinion says: 'abortion is fundamentally different [to other types of law, used as precedents], as both Roe and Casey acknowledged, because it destroys what those decisions called "fetal life" and what the law now before us describes as an "unborn human being"'[6].

The 2nd edition of this book[7] developed out of the official "Opinion of the [Supreme] Court [of America]" being published and the deepening understanding of both what this means for America and the rest of the world. We are now in a better position to see more clearly the inter-

---

[6] Dobbs v Jackson, "Opinion of the Court", p. 6.

[7] My thanks to John C. Lemon for his Foreword to the first edition of this book, *The ABCQ of Conceiving Conception*, 2022, for Dr. Anthony Williams proofreading of both the first and the 2nd edition and for Dr. Sebastian Mahfood's publishing expertise. I wish to thank, too, the following people who made a number of helpful suggestions for the improvement of that first text: Mr. Michal Paszkiewicz; Dr. Dr. Ralph Weimann; and the Rev. Dr. Gareth Leyshon. As regards this 2nd edition, there has been helpful feedback from Dr. Pravin Thevasathan, Dr. Elizabeth Rex, Kiki Latimer and again from Mr. Michal Paszkiewicz; and, more generally, Dr. Sebastian Mahfood's suggestion to go beyond a change of title to a 2nd edition of this book.

relationships between genetic identity, personhood, law, bioethics and, fundamentally, the good of the woman and child, not forgetting the father.

The American debate, then, helps to show the value of a constitution which is framed for the benefit of all. Indeed, it could be argued, that the Constitution could be drawn upon to defend the right to life, precisely because the 14th Amendment says, 'nor shall any State deprive any person of life, liberty or property, without due process of law; nor deny to any person within its jurisdiction the equal protection of the laws'[8]. At the same time, however, as we see with the controversy over what the American Constitution does or does not state, that if laws are going to be developed that benefit the human being, whether threatened with the possibility of abortion, destructive experimentation or freezing, that they will have to be very well formulated, striving as they do to identify what it means to say that a human embryo cannot 'be reduced to the level of an object'.

With respect to the challenges of medicine:

---

[8] https://www.aclu.org/united-states-constitution-11th-and-following-amendments#14.

'Medical knowledge cannot equip us to save the life of every pregnant woman or every baby. When faced with the tragedy of death due to failure of physicians to save a life, the response of a devoted healer is to search for better treatments to prevent such a failure in the future—not to advocate sacrificing one life for another.

Abortions destroy innocent lives in the womb, often in a violent manner, rather than protecting pregnant women at risk, who are often victims along with their babies.

Physicians have a sacred duty to protect lives, especially of the weakest members of society, and offer healing and hope to future generations. The calling of a physician is to be a guardian of health and life, not an agent of death'[9].

---

[9] Abortion: Causing the Death of an Unborn Child", by Dr. Sheila Page, president-elect of the Association of American Physicians and Surgeons, p. 95 of the *Journal of American Physicians and Surgeons*, Volume 27 Number 3, Fall 2022, p. 2 of 2: https://www.jpands.org/vol27no3/page.pdf.

In brief, then, there are so many clustering concerns: justice to all; the nature of the human embryo; the plight of women; forgotten men; the implications of a society in which human beings can be plundered; and, in the end, the help we all need to love, to forgive and to understand one another. But, at the same time, this book appeals to reason, evidence and the benefit of philosophy, the coming together of different faiths and the grace to explain, wherever possible, what will lead to the *truth-in-love*. Moreover, even when people go through all kinds of treatment, including the recognition that they may be carrying an inheritable disorder, it is clear that the humanity of the human embryo speaks to them particularly with the recurrent theme that parents want to avoid abortion; and, in that respect, they seem to be altogether more aware of the humanity of the child than those who "subtract" from the discussion the reality of the human embryo that they are manipulating[10]. Thus human experience goes deeper

---

[10] Violetta Anastasiadou and Joep Geraedts, *et al*, "The interface between assisted reproductive technologies and genetics: technical, social, ethical and legal issues", *European Journal of Human Genetics* (2006) 14, pp. 588–645, p. 599 "Patient's attitudes" and then also p. 604: https://www.academia.edu/

than theoretical discussions about methods of manufacturing and testing of human embryos; but, as we will see subsequently, people who work in the industry of fertilizing the human embryo artificially can discover, too, that they can no longer ignore the humanity of the human embryo and leave.

On the one hand, there are many variations in European bioethical legislation[11], most of which lack a baseline recognition of the protection of the human embryo. On other hand, world-wide[12] legislation that recognizes

---

17345362/The_interface_between_assisted_reproductive_technologies_and_genetics_technical_social_ethical_and_legal_issues?email_work_card=view-paper.

[11] Cf. Anastasiadou and Joep Geraedts, *et al*, pp. 629-641: "Appendix A: Legal Framework" of "The interface between assisted reproductive technologies and genetics' etc.

[12] Cf. "Countries Where Abortion is Illegal 2022": https://worldpopulationreview.com/country-rankings/countries-where-abortion-is-illegal; and note the current controversies where an unestablished so called right to abortion is claimed as established, Stephano Gerrarini, 6th October, 2022, "UN Bureaucrats Claim "Backlash" Against Abortion Rights": https://c-fam.org/friday_fax/un-bureaucrats-claim-backlash-against-abortion-rights/.

the beginning of human life and, by implication, the recognition of the embodied rights of human beings, also needs a new discovery of the relationship between truth, goodness and the gift of life. We live in the hope that the truth about the beginning of human life will establish its own universality; and, at the same time, we hope that while a universal declaration of the natural rights of the human embryo is humanly unlikely, 'nothing is impossible to God' (Lk 1: 37).

However, it is clear that the more we depart from the simple principle, do good and avoid harm, the more there is a proliferation of moral crises and dilemmas; and, therefore, if we are to obtain the good of all, a good as generously given as that of the Creator's gift of existence to each one of us, then it follows that we need to return, to recognize and to rediscover the help of a principled way forward.

Therefore, if this book succeeds in a fraction of its ambition, then perhaps this fraction will be like the crack that starts to let the light in[13] or allows it to spread further.

I pray that each of us will find the starting point that makes further cooperation possible; and, at the same time,

---

[13] Cf. John O'Brien, ofm, *A Love Supreme*.

I pray for those who, for whatever reason, cannot see the full reality of what they are doing and still need to recognize the other as a "brother" or "sister"[14].

REVISITING THE PUBLISHER'S BLURB

Francis Etheredge explores what is involved in grasping the beginning of each of us – but we need the truth to open the heart to love for it to be helpful in humanizing us. If, then, we are equally given the gift of existence, why are we, who are born, the judges of those who will be born or not?

In this 2nd edition, now called *Unfolding a Post-Roe World*, we can see more clearly both the "light" and the "dark" of seeking a stable account of the *American Constitution*, from the point of view of perceiving the right to life of all. But as with Poland's perseverance in the Christian faith in the course of two atheistic types of social

---

[14] This is a distressing article: Lisa Harris, 2nd September, 2008, "Second Trimester Abortion Provision: Breaking the Silence and Changing the Discourse": https://www.tandfonline.com/doi/full/10.1016/S0968-8080%2808%2931396-2.

domination, so with Germany and Italy's protection of the human embryo, in different but complementary ways, we see that a country can have a vocation in world history. Thus the ongoing debate on the 14th Amendment of the American Constitution has assumed a vocational witness on behalf of the whole human family.

Scientists can help us see the reality of the first instant of fertilization being the beginning of human existence – but we need an educated humanity to go beyond the technicalities to the appreciation and development of the relationships that come to exist. So the book explores both an ongoing willingness to adapt our speech to our listeners and the help of God to enable us to go forward in a spirit of forgiveness and understanding of all that is involved.

A Biography and New Foreword by Kiki Latimer

Kiki Latimer has a BA in the Oral Interpretation of Literature, Psychology, & Philosophy from the University of Rhode Island and a Master's in Moral Theology from Holy Apostles College & Seminary in Cromwell, Ct, where she taught Homiletics for seven years until 2020. Her book *Home for the Homily-The Sacred Art of Homiletics* has just been published. She is co-author with Stephen Schwarz of *Philosophy Begins in Wonder* and *Understanding Abortion- From Mixed Feelings to Rational Thought* and has a ten-hour seminar given at Holy Apostles on *Understanding Abortion* available on YouTube. Kiki has also taught courses on Silence & Solitude, as well as Metaphysics, Epistemology, and Virtue Ethics. In addition, she is a children's author with four published books: *Islands of Hope, The WaterFire Duck, BubbleButt,* and *Heal of the Hand.* (www.kikilatimer.com) She is currently the host of *The Catholic Bookworm.*

She was the executive director of *Problem Pregnancy* in Rhode Island from 1985-2005 and is now back as the new director of *Choices Women's Pregnancy Center.* She has four grown children and is grandmother to thirteen grandchildren. She is a 1985 convert to Catholicism and lives with her husband Jim in Hope Valley, RI. Together they have, for 25 years, led a St. Thomas Aquinas *Summa Theologica Wrestling Group* in Providence, RI, over wine and cheese.

THE NEW FOREWORD

The voice of the poet resounds, and the heart of the poet resides not merely in his poetry, but in his prose as well; like a whisper of gold it sneaks into the silent crevices of his prayer, speech, acclamation, dialogue, and teaching moments. So, one finds joyfully, that poet and author Francis Etheredge reveals another moment of a sunset unveiling gold in his *Unfolding a Post-Roe World*, however difficult and sorrowful the topic. In this revisited work of prose, he offers us once again a secret garden view of life and invites us to join him in viewing our world anew.

This is a perfect time for a second edition of Francis's *Unfolding*, for we embark now upon a new world order with Roe undone by the recent Supreme Court Dobbs decision. We currently have a perfect opportunity to revisit old truths in a new light, one in which hindsight is twenty-twenty, the brave new world of the sexual revolution failing to deliver as promised. Francis offers us new ways in which we might look forward with renewed hope and restored vision to the realities of life, love, personhood, and the human family.

> Oh Holy Spirit, Lord of Life and Love, pass among us,
> Reconciling us to one another, to our families, to the life
> We have lived and the love poured out, wisely or not,
> Raising a blessing, like a sunset unveiling gold from heaven;[15]

On the Reality of the Human Person: One in Body and Soul

The great German philosophy professor Dietrich von Hildebrand once remarked as he pointed to a classroom

---

[15] The Poetry excerpts that run, intermittently throughout this Foreword, beginning with "Oh Holy Spirit" are from the prayer called "Visiting", in the book by Francis Etheredge called, *Honest Rust and Gold: A Second Collection of Prose and Poetry*: https://enroutebooksandmedia.com/honestrustandgold/ and the subtitles in the Foreword were agreed with Kiki Latimer.

chair during a course lecture "Compared to the reality of the human person, this chair doesn't even exist." This is a deep call to an examination of the very essence of what it means to be a person. This is a call that the 1973 Roe v Wade Supreme Court decision not only ignored, but dismissed, as unknowable. But we are human persons, and the call to both understand ourselves, as well as know who comprises our flock is radically important human knowledge. For we ourselves, under various and perhaps yet unknown circumstances, could be deemed unworthy of the personhood that we know with certainty by internal perception that we ourselves possess.

An earlier philosopher and theologian, Thomas Aquinas, claimed that the soul is the form of the body, and grace is the form of the soul; grace being our very participation in the life and love of God. Aquinas claims a radical relationship between human body, soul, and grace, so much so, that one without the other is deformed, disintegrated, un-whole. The concept put forth since the Roe decision, that the human body forms in utero absent a soul is a truly hideous concept when put to the test. The Roe decision began the creepy and disturbing descent into separating the concepts of human body and human person, leaving us to conceive of the human animated zombie in utero, devoid of soul/personhood, but living, and growing, a mere human body disposable via abortion due to the supposed absence of soul. The idea that woman's creative power carries not personal human life, but an animated zombie body has been left to run rampant in our culture of death as not an exception, but the norm of cultural thought on the matter of what, rather than who, is destroyed in an abortion. Everyone agrees that something dies in an abortion; but does someone die?

On the Dignity of Women

And what of the status of women? Are they the bearers of new human persons? Is she having a baby? Or is she a womb receptacle, a mere sperm receiver, a personal test tube growing an animated non-personal human body devoid of a soul, devoid of personhood, within herself? Is there anything special about being a woman, a mother, a bearer of new life? Or must equal rights with men strip a woman of that burden, but, ah, also that dignity, that honor, that service to humankind? Is the woman special-beyond-measure, a bearer of a new little child, or is she a mere shell that can be emptied at choice like a wastebasket?

To Choose Life

A possible changing world view following the Dobbs decision suggests that the convenient, violent, culture of death might no longer be the acceptable norm for the future, nor the only choice for the modern woman. Now, even though there will always be some who go for illegal abortions, the majority of women, and men, might once again get the true message that abortion is questionable at best and a grave moral wrong at worst; they may choose life more readily. For when we speak of the choice of abortion, it is best to remember the choice of which one speaks, that being the choice of death rather than life. This is not a mere choice of ice cream flavors at the creamery, chocolate vs vanilla. No, this is a choice of lasting importance concerning a preborn human person's life or a preborn human person's death. And it includes also the moral life of born men and women, for if as Socrates stated: It is worse to inflict moral evil than to suffer it, then to choose and commit abortion involves even graver moral harm to those who are involved in this choosing of death.

Oh Holy Spirit, Lord of Life and Love, pass among us,

> Inspiring new ways to help us with all the challenges
> Of illness, care, finding cures and managing today,
> Refreshing old investigations and breaking new ground;

On the Reasonable Recognition of the Person

Does personhood matter? The Roe decision said it didn't, due to a lack of timely agreement. It gave the hunter in the woods who perceives a shuffle of movement the right to shoot without ascertaining whether or not it is a deer or another hunter; no one knows, there is no consensus on human personhood, so one can go ahead and shoot. But would you? Could you? 60 million people answered yes to these questions since Roe. They shot the deer or rather, the dear one, in the womb.

> But if any dead lie neglected and any left unloved
> Let an angel visit in the moment of need
> Healing, tenderly, the wounds still bleeding,
> And embrace, with an unbelievable blessing,

The heart broken and the life lived almost lost![16]

Just as the Roe decision excluded the entire rational ability to think clearly on the matter of abortion and personhood within the country of the United States, it did so not for our country alone, but for the entire world, to the extent that we often lead the world in many of our views, trends, lifestyles, and culture, for better or for worse; so now the reaction to the Dobbs decision is again upending feelings, understanding, and attempts at rational thought on the sorrowful matter of abortion. Right thought, with as little error as possible, is vital to all human personal life, and the rest of the planet as well, whether it be a matter of biology, sociology, bioethics, medicine, government, philosophy, or theology. Truth matters and we would do well to seek it, lest we perish from a lack of it. U.S. Supreme Court Justice Anthony Kennedy tried to pull the wool over our minds when he blathered idiotically that: "At the heart of liberty is the right to define one's own concept of existence, of meaning, of the universe, and of the mystery of human life." Yes, try to define your or another's human

---

[16] This excerpt is from Part III (II) "The Plight of Passing", in *Honest Rust and Gold*.

existence as a squirrel, or an impersonal blob of tissue, a being that doesn't have moral responsibility and a call to love and be loved; define yourself a being not subject to pain, suffering, or eventually death! See if the universe bows to your definition. The pending shipwreck of that error is immeasurable. Rather, know clearly and irrefutably that at the heart of liberty is knowing truth and acting in accord with it.

In matters where confusion reigns supreme, dialogue, respectful discussion, and deep understandings of the importance of freedom of speech and respect for basic inalienable human rights must be brought to the center ring of the circus of our confusion. As atheist Nathan Hentoff recognized years ago, abortion is not a religious issue but one that can be understood as wrong by the light of natural reason alone. With this in mind, we should all welcome this 2nd edition of Unfolding a Post-Roe World by Francis Etheredge, which seeks to respectfully engage us in a fruitful discussion on the most controversial, sorrowful, and pertinent topic of our times.

> Oh Holy Spirit, Lord of Life and Love, pass among us,

> Renew again the vocation to serve, going beyond our fears
> To the many verges of the world, whether near or far,
> Giving life to dried out hearts and deserted lives

On the Coercive Choice

In my many years as executive director of a crisis pregnancy agency, offering support to thousands of women in difficult situations, it became clear that abortion was rarely ever a true choice, but a matter of coercion by a boyfriend, husband, family member, or friends that either didn't want to offer support or who considered the child an inconvenience at best and a menace at worst. "I have to have an abortion, because if I continue this pregnancy, I will love the baby too much to ever give it up for adoption"[17] became the words repeated over and over again

---

[17] Francis Etheredge: See the subsequent excerpts from Charlie Camosy's marvellous article on the promotion of adoption by Elizabeth Kirk, 7th October, 2022, *The Pillar*, "Supporting adoption in a post-Dobbs America": for what Elizabeth Kirk says of herself and her advocacy of adoption, see

by women who made a clear distinction between the pregnancy which was unwanted and the child who would be dearly loved if allowed to be born.

Many men and woman, born in the era of Roe, have assumed that there is nothing wrong with abortion because, after all, it has been legal. You could shoot whatever rustles in the womb; it might be a deer. Post abortion counseling became necessary to help women, as well as men, both the victims of this lie, deal with the emotional and spiritual devastation that follows abortion at some point in time. One young woman came to me to tell me that in her situation abortion was "the right thing to do" and I listened while she repeated this like a mantra until she fell silent. Then she said "I still talk to the baby every night. I tell him that I am his mommy; I tell him that I love him; I tell him that I'm sorry." He, this little boy, was, after all, a person.

---

https://www.pillarcatholic.com/supporting-adoption-in-a-post-dobbs-america/.

On the Objective Beginning of Personhood

As much as the 1973 legalization of the hideous procedures of the taking of a preborn's life promised escape from an unwanted pregnancy, one cannot escape the consequences of the immoral act of the death-resulting violence of abortion. No matter how well the propagandists parade this objective evil as a subjective good and refer to the bloody ripping apart death procedure of an innocent preborn person as the "gentle emptying of the womb," truth has a way of surfacing in the end. The Dobbs decision is the resurfacing of a truth known since the beginning of time: we must consider that human personhood begins in the womb, that it ought to be nurtured and protected by the mother, and hopefully by the family and structures of the society, and that human persons in the society at large must respectfully and rationally weigh in on this matter of their own children and defend them.

Here, Francis Etheredge takes us systematically through reasons why and ways in which this truth can be unfolded for a deeper holistic understanding of human personhood in its humble and startling beginnings. This involves insight into the subtle, but drastic and desperately important difference between being a person and functioning as a person; only that which has the being of a person can ever begin to function as a person. Thus, getting mired in concepts of functioning such as viability, consciousness, and the ability to feel pain, obscures the underlying more fundamental question of the presence of the being of a person.

It also makes us examine more clearly what it means to look human as our looks may differ radically with race, ethnicity, age, and level of development. Do you look human? If disfigured by disease or accident, can I kill you if I deem you to not look human? This has happened in the past when someone deemed that blacks, due to race, did not look like persons, again in another era of sorrowful history that Jews, due to ethnicity, did not look like persons, and now a claim that the preborn does not look like a person, due primarily to size and level of development. Do we ascertain a pattern here, not of the reality of the person, but rather the pattern of the error of perception of the beholder? Perhaps, like beauty, personhood is not in the eye of the beholder; rather, it is an objective reality outside the beholder. I am a person and the sunset is glorious regardless of another's blindness to these realities! Perhaps the time has come to remove the blindness rather than the realities. The Dobbs decision once again offers us the freedom to consider our own blindness to the personhood of the preborn.

> Let the grace of passing be as water running clear in your hands,
> Our lives as colorful as stones in purest spring water,
> Let the grace of passing be as light in the windows of the soul[18]

Philosophy professor and author Stephen D. Schwarz has identified four characteristics that distinguish the born person from the preborn beginning at conception. These are size, level of development, environment, and degree of dependency. As one moves from conception forward to birth, these are the only four attributes that vary: 1. the being in the womb grows in size, 2. increases the levels of functioning both physically and mentally, 3. the environment varies from the fallopian tube to womb to birth canal to outside the mother, and 4. slowly becomes less dependent, moving toward viability, but is still highly dependent at, and after, being born for at least

---

[18] The next two excerpts are from Part III (II) "The Plight of Passing", in *Honest Rust and Gold.*

several years. It is important to note that none of these four markers are capable of determining personhood, nor have any moral significance. Your size, level of functioning, environment, and degree of dependency cannot make you a person, nor unmake you as a person. Even Dr. Suess got this right: A person's a person, no matter how small! Personhood is not a matter of degree; you either are or are not a person. A non-person cannot develop into a person. Only the radical coming into being at conception of an entirely new being, with unique DNA human coding, from a radical unity of a sperm and egg both at the very end of their life cycles, can come a new being at the beginning of his or her 80 year or more lifespan. Here begins the mystery of being a person; after this we only witness the change and development of the four non-moral characteristics. Once we turn to a subjective concept of personhood as defined by any of these four characteristics, we run the risk that someone in power will judge us unworthy of the title of persons. This might occur for some arbitrary whim, failure to perceive the objective reality of our personhood, a will to power, or a matter of their personal choice, and thus deem us disposable.

Our lives having lost their darkness dripping with
   brightness,
Let the grace of passing be as spring blossoms,
   brightly lit flowers,
Our lives as tulips brimming with your luminous
   presence!

On the Good of the Family

With this in mind, Francis Etheredge offers us a discussion that has immediate importance for the human family of which we are all a member. The Dobbs decision, like the Roe decision, does not take away the presence of a culture in the freefall shambles of a sexual revolution. We still reside in a hook-up culture that has long separated sex from spousal life-long committed love and babies. Therefore, there will still be pregnancies that occur in difficult and sorrowful situations, but we are now more clearly invited once again to handle these with common decency, and care and the nurturing of all involved, once understood as the common good, which includes the unseen voiceless preborn person in utero. This new era allows that women will eventually be seen, once again, as strong, brave, resilient, and capable of doing the right thing in a rough situation. A society that chooses life rather than death is a society once again under the realm of reason, clear thinking, and one that rejects a culture of death and destruction for our youngest people, the future of any society.

> Oh Holy Spirit, Lord of Life and Love, pass among us,
> Reconciling us to one another, to our families, to the life
> We have lived and the love poured out, wisely or not,
> Raising a blessing, like a sunset unveiling gold from heaven.

EXCERPT FROM AN INTERVIEW WITH ELIZABETH KIRK ON PROMOTING ADOPTION IN AMERICA[19]

Elizabeth Kirk: 'Now, I direct the Center for Law and the Human Person at Catholic University's Columbus

---

[19] Excerpt from Charlie Camosy's marvellous article on the promotion of adoption by Elizabeth Kirk, 7th October, 2022, *The Pillar*, "Supporting adoption in a post-Dobbs America": https://www.pillarcatholic.com/supporting-adoption-in-a-post-dobbs-america/.

School of Law, and I teach family law. Adoption law and policy is a particular interest of mine. But, like many people who are drawn to professional work in the space of adoption and foster care, I have a deeply personal connection as well.

I am an adoptive parent, four times over. My husband and I have adopted three children as newborn infants here in the U.S. and have what are considered "open adoptions." Our youngest child was adopted through the foster care system, after a year of us serving as his respite care providers and then his foster parents.

Also, I was the child of an unexpected pregnancy and my mom chose life for me and was a single mom until she married a "just man" (your readers will know the reference!) when I was 3 years old and he became my father through adoption. It is my dad's generous love, his gift of fatherhood, which shaped my views on adoption from an early age – and prepared me to welcome our children.

My experience as a foster mother, an adoptive mother and an adopted child, the state of the current foster care crisis, the horror of abortion, and the urgency with which vulnerable women and children need help compels me to be active on these issues'

Part I

REFOCUSSING THE HEART OF THIS BOOK

The more I became involved in reading, writing, listening, accumulating new material and discussing the Supreme Court of America's judgement, Dobbs v Jackson, the clearer it became to me that this book, even though it was originally called *The ABCQ of Conceiving Conception*, was in fact about the unfolding of both the historical discussion of Roe v Wade and the new judgement.

On the one hand, then, this book discusses both the legal judgements and the ideas that were involved, even taking account of what are clearly redundant ideas that yet persisted right up until Roe v Wade and beyond, invoking ancient theories concerning human conception. Thus, in one sense, this is a book that is very much about different ways of "conceiving conception". On the other hand, with the first edition published, one of the endorsements saw more clearly than the author that the main subject, from which all the other parts derived, as it were, was the opening up of the American debate with, at the time, the expected Supreme Court judgement, Dobbs v Jackson.

This book is still about different ways of understanding human conception, and the need to specify what is clearly now known, but it is also about evaluating, retaining, and discarding, *en route*, the usefulness of the path we have taken to this noble and ethically crucial end. At the same time, given the many misunderstandings, genuinely made, manipulatively employed, or simply based on poor reasoning or lack of evidence, it is also necessary to engage in a kind of parallel discussion about how to communicate the truths concerning human conception which require, as it were, both wide dissemination and a variety of ways of expressing their intelligibility.

If, therefore, the book falls between not being academic enough or not being too consistently simple, it is because of the hope, hopefully not completely unfounded, that we need both: both an intelligent grasp of all that is involved in appreciating the great depths of human conception and the need, manifested in manifold ways, of multiple ways to communicate to the people of today.

What is in a Title? Unfolding a Post-Roe World

Where clarity has begun to exist, it is both necessary to appreciate the new development and, at the same time, to understand the roots of what has been a reigning confusion and, where possible, to provide an answer to the difficulties that prompted a lack of clarity in the first place; it being hoped that the States of America, other countries, and indeed the world, may benefit from reassessing how the beginning of life can be both better understood and expressed in helpful legislation. However, in order to do this, it is necessary to go to the roots of the basic questions of human existence and human rights.

Rights Are integral to Human Existence

What doesn't begin to exist, cannot continue, never mind come to an end. But what has an end, has a beginning. What has a beginning, has begun. What has already begun, had a beginning. Whichever or whatever way we look at it, to exist is different from non-existence; and to begin to exist is as different from non-existence as it is possible to be. Just, then, as human existence is a witness to having had a beginning, why is it difficult to accept that another person's existence is a witness to his or her beginning?

A human right is not a right if it is only for you and not for another. What is for one and not another is not a right but an act of discrimination. An act of discrimination is an injustice. An injustice implies justice. Therefore, ending discrimination against the person from conception, whether about to be aborted, frozen as a "spare" human embryo or experimented upon, requires a just resolution: the recognition that a gift of life is a human equalizer – we all share it and nobody has a right to take it from another as we are all equally in receipt of it. At the same time, a human right is embodied in our very existence – it is not granted, permitted or subject to removal; a human right is both integral to being human and embodied in the very relationships which come into existence with us: of being a brother, sister, daughter, son, grandson, granddaughter, niece, nephew, cousin, sibling and so into the myriad ways we, together, constitute the society in which we live.

There are many natural and true rights which are integral to human nature, to each one of us, beginning with the fact that our existence is a gift: to be conceived is to be conceived-in-relationship to each other and to the human race as a whole. As each natural right pertains to the good of both the individual and the whole human race, there cannot be a conflict between one or more of them; rather, each has its own contribution to make to the whole human good. A right to privacy, then, cannot be a right which implies protecting a wrong[20]. Just as there is no right to deprive an innocent person of life, so there is no

---

[20] This was a right claimed to be integral to permitting abortion; cf. Cf. Tina Beattie, "Catholicism, Choice and Consciousness: A Feminist Theological Perspective on Abortion", *International Journal of Public Theology* 4 (2010) 51-75: https://www.academia.edu/18697950/Catholicism_Choice_and_Consciousness_A_Feminist_Theological_Perspective_on_Abortion?email_work_card=view-paper; on p. 62:

'However, liberal arguments on abortion tend to be informed by this post-Enlightenment concept of the autonomous individual, as was evidenced in the appeal to the constitutional right to privacy in the Roe v. Wade case; a landmark case in the United States that made legal history in the abortion debate, but continues to generate considerable controversy.' So although I agree with her sense of how individualism has corrupted the sense of relationship which she goes on to emphasize, nevertheless, Beattie's relationship between mother (and no mention of a father) is not founded ontologically but in some kind of subsequent recognition by the mother of her relationship to the child (p. 66 etc.) and, as such, is as arbitrary as her other claims e.g. 'no very first instant' (p. 60).

Cf. also the following, very comprehensive grasp of the present situation and its historical antecedents and I only choose one of many points it makes: 'The European Court of Human right to privacy if it is a cloak under which to wrong another. Therefore, there is no human right to abortion. In other words, a right to privacy has a legitimate expression but, like any right, it cannot be used to justify a wrong.

The help we need to understand what is not so "familiar"

> *Imagery is basic to human understanding; and, it could be argued, finding adequate imagery is also an indication of whether or not we have understood our subject and can therefore communicate it easily. Just, then, as moral norms are as integral to human nature as heat is to the flame that expresses it - so the origin of a person's life is when a divine spark ignites it!* [21] This book, then,

---

Rights has held that a mother's right to privacy cannot be interpreted as a right to abortion': "Catholic Medical Association (UK) Submission to the World Medical Association Consultation Regarding the Revised Draft of the International Code of Medical Ethics": http://www.cmq.org.uk/CMQ/2021/Aug/Submission_to_WMA_re_conscience.html.

[21] Cf. St. Paul VI, *Humanae Vitae* (On Human life), 13. Note: A document of the *Catholic Church* takes its title from the opening words of the Latin text and, therefore, the English

like a catechesis, hopes that it will resound in your heart and that you will find your own way to communicate the beginning of each human life to this generation.

This book, then, recognizes that some people have difficulty in recognizing a real beginning and, therefore, takes a closer look at what can be a simple beginning: the growth of plants with a view to helping us to understand human conception[22]. In other words, taking what is more within our immediate experience, we will consider what helps us to understand human conception. This is not an unusual activity; St. Thomas Aquinas says that it is helpful to go from what is familiar to what is unfamiliar; and, as we might imagine, this principle has been used from ancient times. Although the biblical evidence is a different kind of source to biological knowledge, it makes use of what is familiar to help us to understand what is unfamiliar. Thus, the author of the book of Job compares his

---

translation only approximates, in some cases, the original meaning of the Latin.

[22] There are a number of pages in *Conception: An Icon of the Beginning*, which use a variety of examples of irreversible beginnings, e.g. from lighting a match to starting to write etc.

origin with the curdling of cheese: 'Didst thou not pour me out like milk and curdle me like cheese' (Job 10: 10); and David says that God made him an 'unfinished vessel' (Psalm, 139: 16), signifying both the possibility of growth and of being 'earthen vessels' that show forth the power of God (2 Cor 4: 7). But just as Eve has already acknowledged the creative action of God at conception when she said: 'I have gotten a man with the help of God' (Gn 4: 1), so the dogma of the *Immaculate Conception* of Mary helps us to understand the wholeness of the human being at conception – because Mary cannot be wholly holy if body and soul are not one from the first instant of fertilization[23].

At the same time, drawing on the available scientific knowledge of human conception, it can be said that there is a real beginning of the human embryo from the "moment" that the sperm enters the wall of the ovum, and the once inert ovum is now activated, closing around the sperm head and proceeding on an uninterrupted trajectory of human development. Thus, there are simple truths, namely that conception is the beginning of a new

---

[23] This will be more fully explained later in the book; but, for those interested, it is more fully developed in Chapter 5 of *Mary and Bioethics: An Exploration*.

human life; and, in the case of all who have ever existed – each and every one of us has had a beginning. Indeed, the very ongoing life that we live is a living witness to the fact of our beginning. Our development has passed through a number of watersheds: the formation of the human embryo from the union of the father's sperm and the mother's egg; implantation in the mother's womb; ongoing development; birth and the constant growth of the child.

The transmission of human life from one generation to another grounds the recognition that each one of us is equally given the gift of being a person; and, therefore, co-extensive with being a person is being a person in a community of persons: the human race. The rights, then, that pertain to human beings, pertain to human beings from conception; for these rights are integral to the relationship between one human being and another. Human rights, therefore, are inter-relational and express our common humanity: a common humanity which makes each of us a specific human person in the community of peoples.

Finally, then, towards the end of this book, we encounter the social implications of a clearer understanding of human conception: that it can lead to a recognition of the necessity for a widespread, indeed international bioethical law for the benefit of all mankind. Indeed, there needs to be a law that embodies the multiple rights of the nascent human person: the right to the life embodied in the dynamism of every human embryo; the right to completing human development: to the completion of what began unfolding from conception; and the right of ensuring the human integrity of each human being, so that nobody's human identity is compromised by being "mixed" with animal ingredients. Thus, all nascent human beings need to be protected from being "mixed" with "ingredients" from other species, being frozen, experimented upon or discarded. In sum everyone has a right to the integrity of his or her identity and human dignity made as we are 'in the image and likeness of God' (Gen 1: 16f)'[24].

Conception: An Unfolding Content

As some have argued that a person does not necessarily exist from the beginning of human conception, it is necessary to discuss the original meaning of conception. Because we have not always existed, our existence has a

---

[24] John O'Brien, OFM, *Waiting for God: From Trauma to Healing*, Poland: Printed by Amazon, 2015.

Refocussing on the Heart of this Book

beginning: '[a] beginning' is the archaic meaning of "conception"[25]; and conception is also understood to mean "to take in and hold; become pregnant"[26]. Ordinarily, then, we understand by conception that the man's sperm has been received and retained by the woman's ovum, which literally means "egg"[27] and thus she is "with child". In-deed, the word "sperm" comes from the Greek "to sow"[28] seed; and, ordinarily, the sperm is sown where it

---

[25] Con-ception means '[a] beginning' (*The American Heritage College Dictionary*, senior Lexicographer, David A. Jost), p. 228: con-cep-tion: the archaic sense of which is 'A beginning ...'.

[26] Cf. Conception (noun): https://www.etymonline.com/word/conception; and cf. also Conceive (verb): https://www.etymonline.com/word/conceive.

[27] 'ovum (n.) "an egg," in a broad biological sense; "the proper product of an ovary," 1706, from Latin *ōvum* "egg," cognate with Greek *ōon*, Old Norse *egg*, Old English *æg*, from PIE **ōwyo-*, **ōyyo-* "egg," which is perhaps a derivative of the root **awi-* "bird." The proper plural is ova': https://www.etymonline.com/search?q=ovum.

[28] sperm (n.) "male seminal fluid," late 14c., probably from Old French *esperme* "seed, sperm" (13c.) and directly from Late

will contribute to human conception. Interestingly, the expression, *ab ovo*, means, 'from the beginning' or literally from the egg[29]. We can see, then, that these ancient expressions point to a simple understanding of "a beginning", which is from the sowing of a seed which has been taken in and held by the egg that, instantly, changes from being an egg to being a human embryonic child. Thus the natural world, from which these ideas were drawn and from which they come together to develop an insight about a beginning, is the wider context upon which our understanding draws; indeed, as we know, to conceive an

---

Latin *sperma* "seed, semen," from Greek *sperma* "the seed of plants, also of animals," literally "that which is sown," from *speirein* "to sow, scatter," from PIE **sper-mn-*, from root **sper-* "to spread, to sow" (see sparse): https://www.etymonline.com/search?q=sperm.

[29] 'ab ovo "from the beginning," Latin, literally "from the egg," from *ab* "from, away from" (see ab-) + ablative of *ovum* "egg" (see ovum). The expression is said to refer to the Roman custom of beginning the meal with eggs, as also in the expression *ab ovo usque ad mala*, "from the egg to the apples" (Horace), hence "from the beginning to the end" (compare early 20c. *soup to nuts*)': https://www.etymonline.com/search?q=ovum.

idea is to make a beginning which, in the course of developing it, has whole history implicit in it (cf. Psalm 139).

Conception, more fully understood, is therefore through the husband and wife's gift of themselves to each other and, just as love goes out to others, so beyond the gift of this self-giving is the possibility of a child. *But just as God regards the good of human life as distinct from the fallen nature of man through which we come to exist (cf. Gn 3-4), so God regards the good of human life even if the child is conceived through rape, in a glass dish or by any other means than through the spousal union of husband and wife.*

> 'St. Thomas explains that ... "God allows evils to be done in order to draw forth some greater good. Thus St. Paul says, 'Where sin increased, grace abounded all the more' (Rom 5: 20); and the *Exultet* sings, 'O happy fault, ... which gained for us so great a Redeemer'"'[30].

---

[30] St. Thomas Aquinas, *Summa theologiae*, 3, 1, 3 and 3; cf. CCC, 412, quoted in *The Navarre Bible: Pentateuch*, Dublin: Four Courts Press, 1999, p. 55.

If, then, the father's gift of what transmits human life, his sperm, becomes one with the mother's gift of giving the ready-to-receive "egg", then conception is expressed in the moment of the egg enclosing the sperm. More precisely, then, the first instant of fertilization is the first instant that the sperm of the man animates the ovum – because up to this point the mitochondria, the energy-producing cells in the ovum, are inactive. In other words, the closing of what is now the embryonic wall is the first act of a totally new human being; and, ordinarily, the embryonic child is a newly formed and independently growing, indwelling person, both con-ceived within the mother and "im-planted" in her womb.

Conception Expresses both Uniqueness and Relationship

On the one hand, the inheritance of our common humanity is through the reciprocal self-giving of the parents; but, on the other hand, the nature of being a person exceeds what is transmitted by the parents. Human development, then, shows forth the implicit dynamism of the person conceived. Beginning, then, with whether the child is male or female, development unfolds all that is entailed in being the person each of us is conceived to be. The natural unfolding of the child's humanity is expressed in the physical, psychological and social development characteristic of each one of us. Our human being is inseparably physical, social and psychological; and, therefore, as the child is conceived so begins, in that same instant, the integral being of the human person: each of whom is a being-in-relationship. There is an implicit social nature to the human embryo in that not only does he or she come to exist in virtue of what she or he has received from the mother and the father – but also because the very presence of the child in the womb expresses a particular relationship to the mother.

There is a kind of biological dialogue in which there are subtle exchanges between mother and child. There are at least two kinds of biological mother-child exchanges. On the one hand, 'this biochemical dialogue between mother, [fallopian] tube and child allows the embryo to move forward at the right speed to be able to access the uterus at the precise time for its proper implantation'[31]. On the

---

[31] Francis Etheredge, *Conception: An Icon of the Beginning*, Chapter 5, Part II, p. 500, prepared by two experts, the first of whom once assisted the *Pontifical Council for Life*, entitled: "The Biological Status of the early Human Embryo: When does

other hand, there is a relationship between what the mother eats and what the fetus senses: 'by swallowing and inhaling the amniotic fluid, a fetus can sense the flavors of food eaten by its mother' .... [and so] 'by analyzing their facial reactions, we present direct, novel evidence that fetuses can discriminate different flavors in amniotic fluid'[32]. Furthermore, as the mother's awareness of the child grows, literally, with the growing of the child, so the mother-baby dialogue is also psychological: the mother looks forward to meeting her child[33], as well as the mother's whole awareness of the child's movements and

---

Human Life Begin? A Paper by Professors Justo Aznar and Julio Tudela": https://enroutebooksandmedia.com/conception/. Furthermore, on p. 500, there is footnote 704, which comes at the end of this quotation, and cites the following work: 'Tudela J, Estelle R, Aznar J. Maternal-foetal immunity: an admirable design in favour of life. Medicina e Morale. 2014; 5: 833-45.

[32] Beyza Ustun *et al*, "Flavor Sensing in Utero and Emerging Discriminative Behaviors in the Human Fetus": https://journals.sagepub.com/doi/10.1177/09567976221105460.

[33] Cf. Francis Etheredge, "The Mysterious Instant of Conception", *National Catholic Bioethics Quarterly* of America, Vol. 12, Autumn 2012, No. 3, p. 422 of pp. 421-430.

reactions. Let us note, too, Elizabeth's awareness of John the Baptist's response to Mary being pregnant with Jesus (cf. Lk 1: 44).

The child exists, at once in the present, in relationship to parents, relatives and any number of other people. Indeed, as the mother looks forward to the birth of her child, so she looks forward to meeting the child she nurtures; and, similarly, as the father witnesses the unfolding of his child's development, so he begins, too, to look forward to the birth of their child. The embryonic unfolding entails psychological development and, therefore, as the child responds to the voice of the parents and, from birth, to being fed and to smiling and to amusing sounds, so the whole world of relationships multiplies with grandparents and the many people with whom they all come into contact and with whom they are all connected.

As the child begins to put into words his or her experience of life, however simply, so we can see that our inner-life, our consciousness of our own experience of life, develops through our communication with each other. Indeed, we can even call it a law of psychological development that what is expressed contributes to our human development and what is inhibited or repressed obstructs it: communication of the self enables the differentiation of the self through the relationships that help us to know ourselves through knowing each other.

Just, then, as there are the diverse laws of physical, psychological and spiritual development, so the soul is the principle of unity in the human being; and just as laws regulate the fulfilment of coming to exist, and exist as inner determinants of growth, so there is the Divine Lawgiver who brings the whole person to exist, one in body and soul. In other words, the invisible act of God which brings to exist the embodied human person, one in body and soul[34], is shown forth through all that unfolds from conception. Physical and psychological development, interconnected as they are, shows forth the primordial unity of the person: that the nature of the person transcends matter in that the human body expresses a human soul[35]; and, at the same time, as the relationship to the parents develops, so the spiritual nature of the whole person who

---

[34] Cf. *Gaudium et Spes*, (Joy and Hope) 14.

[35] Cf. Pope St. John Paul II, *Familiaris Consortio* (The Family Communion), 11; *Evangelium Vitae* (The Gospel of Life), 60.; and *Veritatis Splendor* (The Splendor of the Truth), 58-50.

exists[36], by virtue of being created by God, is called to a relationship with Him.

Questions

There are many questions which can surround our understanding of what is going on; indeed, these questions can be like a wall of brambles, a fire wall or a dense fume-fog, which prevent us from recognizing the simple truth of human beginning. Are we justified in describing the father's sperm as seed? Are we justified in describing the mother's ovum as an inert egg? Does the child's life begin before conception? Is the sperm or the egg the beginning of the life of the child? Does the child's life begin after the fact of conception? Does the child's life begin at birth? Does the child's life begin when the child becomes conscious?

The father's sperm, like a small dynamo, seeks the enclosure which will enable it to energise the existence of a child; for, while male and female contributions are equal but different, it is God who gives the existence of the

---

[36] *The Catechism of the Catholic Church*, 28, 44-45, 327, 355, 362-368 etc.

person-as-a-whole. The father's sperm contributes to the existence of a child, just as the mother's egg does so, in that as soon as the mother's egg encloses the sperm, shedding what is unnecessary, the sperm drives the changes of the immeasurably larger egg which has now become a single celled human child: a human embryo. A child is not the father's sperm nor is the child the mother's egg; but a child is the coming together of the father's sperm and the mother's egg. Even in the case of twins, which have originated from the same conception, when the early embryo divides and there are two children, then each child has a beginning; one from the beginning of conception and one from the point of being divided from the original embryonic child's development. In other words, where the child's embryonic body lives, there the child is developing. As regards the possibility that, in the case of the splitting of a single embryo, forming two human embryos, there is the death of the first child and the beginning of two more, it presupposes that there can be a death without a dead body. In other words, so integral is the body of a person to being a human person, that there cannot be a human person without a human body; and, therefore, if there were to be the death of a first child, giving rise to a second and a third, then there would have to be the evidence of the death of the first child's bodily expression. In view, then, of twins arising out of the division of a single human embryo, the very continuity of there being two lives gives evidence that one life has continued and a second has arisen from it.

Given the existence of twins arising from a single human embryo and, on occasion, more than twins, there is the problem of "sampling" the human embryo to test it for one kind of disorder or another. There are at least two if not three problems with this practice: a disregard of the original and significant question of the identity, dignity and integrity of the human embryo; damage to the original child in view to a non-therapeutic intervention; and the risk that "sampling" a human embryo in that at an early stage of embryonic development all the cells, if they separate, can possibly become siblings[37].

---

[37] Cf. Violetta Anastasiadou and Joep Geraedts, *et al*, "The interface between assisted reproductive technologies and genetics: technical, social, ethical and legal issues", *European Journal of Human Genetics* (2006) 14, pp. 588–645, p. 595: 'All the cells in a human embryo, at four- or eight-cell stage, are believed by many to be totipotent, that is, none of the cells is yet

More generally, however, a person might argue that because the human embryo does not look like a mini-adult then the human embryo is not a human person: that what does not look like a human being is not a human being. But does the child look the same at the end of life, if he lives a long life, with greying hairs, perhaps less teeth, scars, maybe glasses and walking stick, wrinkles, moustache, and beard? In other words, even from birth the process of growing up goes on and changes the outward appearance of the boy becoming a man, a man going into middle-age and a middle-aged man becoming older and even elderly. Even as we go from childhood to adolescence there are changes that might make the child unrecognizable as, for example, the young child who is scarcely more than a toddler growing to beyond six feet tall. Ordinarily, then, growth entails change and change expresses what is constantly present, whether we are male or female; and, therefore, a man manifests what is there from the

---

committed to a specific developmental path.' Spontaneous twinning demonstrates this possibility.

beginning and what is there from the beginning entails what develops into a woman.

Is My Body Mine?

But, says a woman, my body is mine?

Our bodies express each one of us; and, as such, my body is essential to the gift I have received in being given to be me; and, from the point of view of the unity of body and soul, it is true that the body expresses the person-as-gift: he or she gives what is to be given. But, from the point of view of the whole gift of the person, what is received-as-gift is to be given-as-gift; and, therefore, the gift of male and female is integral to the person-as-gift[38]. Within the context, then, of the giving and receiving the person-as-gift is the conception of the gift of the child. In other words, being human is an ontological gift: a gift of being-as-gift: that human existence, indeed all that comes into existence, comes into existence as gift. The child, then, comes to exist with a view to being given the wholeness of

---

[38] These expressions are indebted to the Theology of the Body of St. John Paul II.

human personhood; and, therefore, just as a present leaves the giver to be given, so the child is conceived to be given into the hands of his or her self.

I am born, so am I conceived to be born; and, if there is a problem with my development, nevertheless, the orientation of the whole process is to birth. Indeed, in order to accept what is no longer an expression of her own body, the mother's immune system has to let it accept the indwelling of the child; and, therefore, it is clear that the woman bears the child conceived – but bearing a child is a relational process: a relation between mother and child and father. The problem with the perception that 'my body is mine' is that there is no sense of being a gift-in-relationship: of having received a gift of existence and, in a sense, out of gratitude, being willing to contribute to another person's gift of existence. Just, then, as the sperm left my father and the egg was released within my mother, so being conceived is *en route* to being born. The child, then, is not an organ of the mother. The child's placenta, which sits within the womb and, ordinarily, comes away with the birth of the child, helps the mother accommodate the presence of the child; it acts as an "exchange", governing what the mother gives the child and what the child excretes. The placenta, then, is a temporary organ and, as such, is a witness to the child dwelling, temporarily, in the mother; and, indeed, the very growth of the child testifies to the independence of the child from the mother for, ultimately, the growth of the child is towards being born, fully formed, with a view to the ongoing growth and development of his or her life. What is more, fetal cells pass through the placenta and, it seems, have a beneficial effect on the mother: "'They can go to the liver and become liver cells, or go into the heart and become muscle cells," Nelson says. Fetal cells can even cross the blood-brain barrier and turn into neurons'[39]. Clearly, this contrasts with the destructive "stealing" of the embryonic stem cells which destroy the embryonic human being and is, as it were, a kind of surplus from which the mother benefits. Moreover, the whole reciprocal benefit of the presence of the embryo is only owing to the presence of what was not there before; and, therefore, both this and any kind of

---

[39] Cf. also an article on the transfer of fetal and maternal cells, one to the other, through the placenta: Michaeleen Doucleff, October 26th, 2015, "Fetal Cells May Protect Mom from Disease Long After the Baby's Born": https://www.npr.org/sections/health-shots/2015/10/26/ 449966350/fetal-cells-may-protect-mom-from-disease-long-after-the-babys-born.

implanting of human embryos testifies, incontrovertibly, to the human embryo as a human subject in his or her own right[40].

What we need are helps to the imagination; for, in general, while many of us will have seen pictures of early embryonic human life, the birth of a child, or babies, there are still many difficulties recognizing what we see as "being one of us". In a word, however, each one of us is a witness to the fact that we are not a bodily part of either the man who contributed to our existence, or the woman who bore us; rather, we have been given the help that belongs to men and women to give that is given to exist.

This Book

We need, then, to draw on a variety of comparisons to help us to understand the mystery of human conception; and, therefore, human conception is put in the context of the natural imagery which helps us to know what is

---

[40] Cf. the whole problem of IVF and pre-implantation diagnosis: Anastasiadou and Joep Geraedts, *et al*, "The interface between assisted reproductive technologies and genetics: technical, social, ethical and legal issues".

Refocussing on the Heart of this Book 59

happening. We need help to recognize what we see; and, recognizing what we see entails understanding it. One of the principal helps to understanding is imagining comparisons which make what is unfamiliar more familiar. However, imagination can both take us away from what exists or it can lead us to it; and, indeed, imagination itself tells us about ourselves too.

But if we begin thinking all things are possible and that there is no limit to what we can imagine then we have already begun to envisage possibilities. We can imagine, then, that what we are is something semi-insubstantial, like a computer program, which can be downloaded onto a variety of different computers[41], without altering what it is. While, however, this use of the imagination has worked very well for computers and the programs which run them, what it consists in is a compatibility between the program and the computer: a set of instructions, however adaptable, and a machine that can recognize those instructions and action them. There is, as it were, a hand and

---

[41] Cf. Stephen Law, "Would You Want To Live Forever": https://www.linkedin.com/pulse/would-you-want-live-forever-stephen-law-phd/?trackingId=ONAfacRLSuyuKtT5%2FfDzqw%3D%3D.

glove relationship between the computer and the program. In the case of human beings, the person is not a program, "life", and a "body", that life activates; rather, the human person is that totality of "human life" and "body". While the body lives, there is the life of the person; and, according to the kind of activities that are possible we understand ourselves more clearly: a writer writes; a singer sings; a talent, in other words, shows itself in an activity.

Thus "imagination" is that by which we envisage many paths, either as goals of understanding or goals of action; and, while we imagine these different possibilities, there is a dialogue with ourselves and what exists as to which one is really possible for us – either as individuals or as members of the human race. Walking or travelling is about as universal as it is possible to be in that if we are disabled, and cannot walk, we can still travel by various means to a destination; and, equally, there can be a wide variety of reasons why we want to go – all of which involve choices of one kind or another. Choosing goals is both a common characteristic of human beings and distinguishes us, too, by that use of the imagination whereby we bring to exist what did not exist. Recognizing our goals as real and good, our methods too, entails feedback from reality, which enables us to reason to the truth and, if necessary, modifying our goals or methods, and the whole toing and froing which this involves as we go from task to task. Imagination, choice of goals and the use of reason are all factors which express being a person.

The imagination, however, with which this book is concerned, is what we use when we want to understand what is unfamiliar on the basis of what is more familiar to us. No doubt, as this book progresses, it becomes more difficult to understand because it is necessary to draw upon embryology, philosophy and theology to help us; and, therefore, we need to begin simply, remembering that simplicity, when the relationship between law and human rights make it more complicated.

Five Parts

What seems, then, to be an unusual and unfamiliar approach to the subject of human conception, of seeing how we have and do conceive of conception, as we will see, is perhaps more characteristic of how people have tried to understand our origin than we may have first realized. But, in the light of modern science, we have become clearer about when and where our natural capacity for making comparisons or analogies can be improved or abandoned for the simple truth of what we now know to be true of our beginning. At the same time, however, we do need to be able to use our imagination to help us to grasp, by comparisons, what is in reality both a visible and an invisible reality: the instant of coming to be a whole human being, one in body and soul.

This book, then, is divided into five parts; and, it is hoped, while these parts address different aspects of the whole, that they are complementary. This book seeks to stimulate the founding of an internationally binding charter which ensures the natural integrity and the completing human development of everyone, from conception until natural death.

Finally, the book cannot be exhaustive, nor is it intended to be; rather, it is hoped to be a start, even if its starting point is not zero; but, nevertheless, it is written with a view to helping a person become more familiar with the ordinary extraordinary event of human conception. Indeed, there are different ways of "conceiving conception", as we shall see, which entail making many "comparisons", both historical and in the present, that either help or hinder becoming clearer about the reality to be investigated: the beginning of each one of us. Just, then, as there is an embryological science which sets out to identify what is known about the reality of human beginning there is a need, too, to find ways to connect this with our ordinary experience and so to be able to communicate the truth of a real beginning more and more widely. Thus, the more widely the truth of an actual, real, irreversible beginning to each one of us is capable of being communicated the greater the possibility of discovering a legal expression of who, in the end, exists in relationship to all of us from conception onwards.

Part II

The Teaching of Experience: Law; Analogies; and Gardening

COMPRISING CHAPTERS ONE AND TWO

This book began, as it were, with the third and more difficult discussion on the need for bioethical law, drawing on the many problematic aspects of the American Supreme Court's ruling in *Roe v Wade* which, nevertheless, helped to set out what needs to be understood and addressed in good bioethical law: law that will both identify and protect the human person from conception onwards. In due course, as the American Supreme Court's, subsequent judgement in Dobbs v Jackson, was leaked, so the discussion developed until, seeing the actual judgement promulgated by the court led to this 2nd edition of the book. Thus the process and the actual judgements are both informative about judicial processes and instructive as to what judges consider as relevant to a particular case; however, in addition, the law itself is a powerful teaching instrument – directing the learned and the unlearned alike, making a path to the truth more direct or obscure and hidden amidst a variety of claims. Unravelling these judgements, past and present, entails the range and content of these chapters.

The Chapters in Part II, however, draw out the need to explore different methods of teaching the truth about the beginning of life; and, indeed, it is a matter of each person adapting their own experience to suit this purpose. On the one hand, there is the embryological evidence which we will examine in the course of this book; but, on the other hand, there is the challenge of understanding an irrevocable beginning. Therefore, while this introduction will give a variety of examples of a beginning, the chapters themselves draw particularly on different aspects of gardening. It may be, however, depending on the ambience of your culture, that it is better to have a different starting point. Either way, we need both the relevant science and ways to make this science widely accessible. One reason for these examples is that some people have difficulty in getting beyond the obvious: that the very beginning of human life is almost completely unrecognizable as the life of a person.

Consider, then, the artist who gathers materials in order to paint a portrait; at the start of the process of painting, which may be more like drawing in colour, a few lines may not look like the subject at all but, as the work progresses, so the likeness to the person emerges more clearly. Nevertheless, right from the first stroke, the artist is aiming, as it were, at the portrait of a person. Alternatively, the ingredients of a cake, a Victorian sponge, when laid out, include butter, flour, sugar and eggs; but, even when it is all mixed together, it looks more like a creamy paste than a cake until, heated in the oven, it irreversibly changes into a recognizable cake. Nevertheless, right from the beginning, the baker is making a sponge cake. A washing machine design, laid out on a piece of paper, enlarged to the point where the intricate circuits are all visible but the whole picture is too large to take in at a glance, is still a design on the basis of which an automatic washing machine is going to be made; and, indeed, the first few parts, whether frame and wheels, circuits or mechanisms, may not resemble the finished item at all. In other words, in each case we are addressing the discrepancy between what an object is at the very beginning of being made and yet, at the end, it is a recognizable picture, cake or washing machine. This is the kind of problem encountered at the beginning of a person's life: that the beginning of being a person is not immediately recognizable as the boy or girl who is, slowly, irreversibly, in the process of being shown to be present from the beginning. Let us use what imagery and comparisons that we can, then, to "grow" a more widespread understanding of the beginning of human personhood.

Even the lighting of a candle, for example, can literally ignite the thought that there was an irreversible moment when the wick was not alight and now it is; or, if the light is not passed, then the ignition of the candle has not occurred. But, on the other hand, if the candle is lit and then goes out there is the equivalent of a miscarriage. Alternatively, knowing that there are innumerable mitochondria, or power generating units, present in the woman's inert egg, it is clear that lighting the candle is passing a light to what can sustain it owing to the presence of the candle's substance; so, as with egg from the woman's ovary, once "ignited" by the sperm, is a dynamically irreversible process of being the human being present from the first instant of fertilization. At the same time, without the light being passed, the candle remains inert, just as the egg remains inert and passes away if not fertilized. Alternatively, however, it could be that the imagery that arises out of technological devices, like the sending of the text becoming, as it goes, irreversible and, even if the device is destroyed, the text remains on a server somewhere in the "world"; existence, then, is an irreversible condition of what exists and, therefore, once the person has come into existence, his or her existence is now an existence-in-relationship: the boy or girl is a child of parents.

In what follows there will be a somewhat extended discussion of the beginning and development of plants in that these are both organic processes, involving life and, at the same time, they have been used from ancient times to describe the beginning of human life. On the one hand, then, these are helpful considerations that may well have influenced early philosophers first attempts to understand the actual process of human development. On the other hand, it may help to distinguish, very clearly, what is useful in these comparisons from what clearly does not apply to human beings. In other words, without realizing it, all the time I have been growing vegetables in the garden I have been thinking about the consistency of what grows: that just as a germinating tomato seed becomes a tomato plant, bearing tomatoes, so children are the fruit of human parents – even if those parents are unknown to the child or in some way indirectly present as with taking the "parts" of a beginning and putting them together artificially, as with cloning.

Even if there are many imperfections with this discussion, there is a definite value in our everyday experience and, therefore, I encourage us all to consider the value of the everyday arguments for the first instant of human conception and the bioethical questions which are involved.

In the end, however, a child is not a picture, a cake, an object or a plant – but a person: a being-in-relationship; for, from the first instant of fertilization, he or she is a child of his or her parents; and, at the same time, if there is no child then there are no parents: a child is always a sign of the relationships between the generations. What is more, as with all these examples, there is the more or less hidden action of the presence of the painter, the cook, the designer, the candle-stick maker or the gardener. In other words, God is always present at the precise moment of the first instant of fertilization, bringing to exist the whole person, wholly present, from the first instant of his or her existence. At the same time, whether it is the paint brush, the mixing bowl, the plant pot or the tools, there is always what was needed to bring about the picture, the cake or the plant – but which is not an integral part of it. Similarly, with human development, there is what helped to bring about the existence of the child but which is not a part of the person who comes to exist; but, in the case of the child, there is the enduring nature of relationships – the child is a being-in-relationship! Again, however, even if it is a pale resemblance to the reality of the relationship between God and giving existence, between parent and child, an artist paints, a designer makes objects, a gardener grows plants.

In other words, we live in a world constituted by relationships of all kinds; and, therefore, to deny this, to act against it, is to deny the dynamic expression of our very being: that to exist is to be a being-in-relationship. It is hoped, then, that this slim book contributes in some small way to opening our eyes to the relationships integral to our life and our salvation; for God, who acts at our beginning, acts with a view to our whole life and end – for He wants no one to be lost (cf. 2 Peter 3: 9).

Five examples of the image as counter argument

Let us take, then, an argument that is advanced, as it were, by way of suggesting that we do not know what is present at the beginning of a human life: that 'not all of the early cells will develop into parts of the embryo. Some will build the placenta but which ones is not decided before 14 or 15 days of the early embryo's development, when some of the middle cells begin to differentiate'[42].

---

[42] Anna Alichniewicz and Monika Michalowska, "Medicine of the Beginning of Life. Bioethical and Philosophical Arguments in the ART Debate", Warszawa, 2019, p. 18:

Just, however, as we do not confuse a diver with the breathing equipment that is needed to survive under water, we do not need to confuse the human embryo with the amniotic sac and the placenta that makes the infant's survival during growth in the womb possible. The fact that, at an early stage, it is almost impossible to determine which cells are the human embryo and which are the amniotic sac does not detract from the reality that a child, like a diver with breathing gear, needs what is provided to thrive in the womb.

What is the difference, then, between cutting or removing the diver's breathing equipment and a non-therapeutic intervention on the human embryo, especially when it is unclear which part is the breathing equipment and which is the child?

https://www.academia.edu/42104311/Medicine_of_the_Beginning_of_Life_Bioethical_and_Philosophical_Arguments_in_the_ART_Debate?email_work_card=view-paper.

Second: The 'research embryos, when there is no intention of implantation, do not have the potentiality to become a developed human being'[43].

However, a human embryo designated for research is no less a human embryo; and, therefore, determining whether a specific human embryo is going to be implanted or "experimented upon" does not determine the identity of the embryo-as-human. In other words, if another person keeps the diver under water to examine how long it will take him or her to die – it is still a human being who will die.

Third: With respect to preimplantation diagnosis, 'It could be concluded that to date there has been no empirical evidence to support the view that children born with PGD [Preimplantation Genetic Diagnosis] are at risk of side effects and that the procedure itself will

---

[43] Anna Alichniewicz and Monika Michalowska, "Medicine of the Beginning of Life. Bioethical and Philosophical Arguments in the ART Debate", Warszawa, 2019, p. 31.

have a detrimental impact on their health in the future'[44].

In this third example, the discussion about the ethics of preimplantation diagnosis or, for that matter, any diagnosis of fetal conditions, could already imply a "product mentality": a "test and discard" mentality. This is true, whether it is the parent, doctor or both who decide on which human embryo to transfer to the woman and which to reject[45]. While it is possible that the test, presuming it is accurate, could be of benefit for the intrauterine or postnatal care of the child – it is often, as I say, part of a different mentality: that of a "test and discard" mentality.

But what is the impact on the parent's relationship to the child when there is a "product" mentality? On the one hand, a defective toy or garment can be returned to the shop and exchanged for non-defective goods. On the other hand, in the case of a child, the human being is an

---

[44] Chapter 3, p. 77 of Alichniewicz and Monika Michalowska's, "Medicine of the Beginning of Life." Etc.

[45] Anastasiadou and Joep Geraedts, *et al*, pp. 603-604 of "The interface between assisted reproductive technologies and genetics" etc.

integral person, an incarnate soul (cf. St. John Paul II, *Familiaris Consortio*, 11): an outwardly expressed personal identity. The person, then, is non-returnable. If however, the person is discarded, what is the future dialogue with others going to be – if, that is, there is any honesty: "You are the child we kept because you were not defective – but your sibling(s) …".

'The problem with follow-up studies [after PGD] is, however, that only a minority of parents want to inform their children about the use of ART'[46] [Assisted Reproductive Technology][47].

In other words, there is a whole wealth of hidden suffering when it comes to admitting the full reality of how a child comes to exist if there has been such a production process involved.

Fourth: To claim that 'Since embryos are not autonomous beings, and as such do not have any goals or

---

[46] *Ibid*, p. 609 but also see 611-612, and see 619.

[47] *Ibid*, "Glossary", p. 644.

interests ...'[48] is to express a profound ignorance of the embryological reality that implies an "abstracting" mentality in the world of research: an abstracting from the reality of human relationships and the reality of embryological development. This is because embryological development is totally 'autonomous', and goal orientated, expressing the child's 'interests' in the course of the psychological maturity that unfolds on the basis of biological development. In other words, to claim that an arrow that has been prevented from reaching its target is defective – when it was prevented from reaching its target by a screen being thrown up between being fired and arriving at its destination is like blaming the pedestrian for being unable to get out of the way of a reckless driver.

Finally: Deciding on an outcome for children without respect for the truth-in-love is like ignoring the conscience-as-compass on the journey through ethical choices; moreover, choosing intelligence, height or any other attribute over the person is like attaching a large

---

[48] Chapter 3, p. 86 of Alichniewicz and Monika Michalowska's, "Medicine of the Beginning of Life." Etc.

magnet to the compass, without realizing how it distorts the direction to go in.

CHAPTER ONE

AN IMAGINATIVE USE OF GARDENING AND PLANT LIFE

'The seed is actually an "embryo" that simply needs "gestation" - food, water, warmth - in order to grow. So the earth is more like the "womb" where the new embryonic plant (with its own unique DNA) can "implant" and grow[49].

'So many plants reproduce by fertilization - pollination - and the fruit is the result'[50].

Before discussing the experience of gardening, with a view to how it can be a source of imagery and experience for understanding the ways of nature and, in particular, different aspects of conception, it may help to establish the

---

[49] Email correspondence, a good summary of the embryonic plant by Dr. Elizabeth Rex, 03/09/2022.

[50] Email correspondence, again a good general comment by Dr. Elizabeth Rex, 03/09/2022.

difference between the science of gardening and the use of gardening in a catechesis on the beginning of life.

Part I: The Science of Tomato Plant Reproduction

Beginning, then, with the summary of the reality of plant seeds, that they are embryonic plants, that is, plants that are ready to grow into what they are, we can see that there is a justified criticism of comparing the plant seed to sperm and the earth to an egg. This is because the plant seed, as has been well said, is already an embryonic plant and, when planted in the soil, is more comparable to the "implantation" of the human embryo in the lining of the womb. Whereas, in the case of the human sperm and egg, each are distinct, but complementary contributions to the transmission of human life; but, sperm and egg, when they come together with the action of God, bring about the existence of the human embryo.

The technical name for a 'flowering, fruit-bearing plant', such as a tomato plant, is 'angiosperm', which comes from 'a couple of Greek words where *angeion* stands for "vessel" [or receptacle] and *sperma* means "seed"[51]; and, as such, these plants obtain their name from the fact that there is both sperm and ovary in the flowering plant. On the one hand, there is the sperm or pollen, which will fertilize the ovary where the new tomato seed will form. On the other hand, the tomato plant begins the process of forming the tomato seed, when the pollen comes from the male part of the flower called the stamen and descends to the ovary, the female part of the flower, it meets the ovule, which, on fertilization, will form the new tomato seed[52]. The male part of the plant produces the pollen and it is from the pollen that there comes the sperm which is what actually fertilizes the ovule and produces the tomato seed[53]. The ovary, which contains the ovules, which become the tomato seeds, will then swell and become the flesh of the tomato: the soft red flesh of the tomato, whether large or small, which is then picked off the tomato plant and is what is generally served to eat.

---

[51] "Angiosperm": https://www.biologyonline.com/dictionary/angiosperm.

[52] "Ovule": https://biologydictionary.net/ovule/.

[53] "Organismal Biology": https://organismalbio.biosci.gatech.edu/growth-and-reproduction/plant-reproduction/.

In general, because there is a male part of the plant, producing pollen-sperm, and a female part of the plant, called the ovule, in the ovary, this type of plant reproduction is called pollination, which is a type of sexual reproduction; and, as such, it could be helpfully used to explain fertilization: which is what happens when the human sperm encounters the human egg. Just as the pollen producing sperm fertilizes the ovules in the ovary, so a male sperm fertilizes the female's egg.

Part II: Teaching Through the Everyday Experience of Growing Vegetables

There is a challenge, as it were, in the early "moments" of conception. Just as the tomato seed is sown in the soil, and it is a couple of weeks before the first two leaves come up, while the roots have been going down, so the pregnant woman does not always know that a child has been conceived and, in the quiet of the womb, has implanted and continues to develop. In other words, we need a variety of explanations which help to educate us all about the first instant of a human embryo's existence.

It is more than possible that a careful use of the sexual reproduction of some plants, such as tomato plants, can be used to help us to understand human procreation; however, as a parent, this can be a part of what is communicated to young people using a variety of imagery. So, for example, the mutual giving of husband and wife is a bit like the giving and receiving of a present in marriage which, until the celebration of marriage, remains wrapped – even if in reality that wrapping can be very torn by the time of marriage or maybe had to be repackaged through the forgiveness of sins. In another conversation about the early loss of a sibling through miscarriage, one of my daughters and I were thinking about this when she declared that Hilary, whom we had named, must have been the size of a dust particle; indeed, I seem to remember that I was driving and it was sunny and maybe a dust particle had caught the light and it lit up with the "image" of Hilary. This wider use of imagery, then, is the context in which to understand the following use of the experience of gardening. As, often as not, we need what is familiar to help us to understand what is unfamiliar; however, we also need what is familiar to be so. What is familiar is like a steppingstone to what is unfamiliar.

Therefore, while in one sense computers and computer speak are more familiar to most people than gardening, gardening needs to become more familiar again to help people to understand the action of nature in the growth of plants and, ultimately, human beings. So this chapter is both advocating a return to growing plants from seeds and the use of that imagery and experience to help us to understand the mystery of human life. Different plants, as we shall see, are helpful in different ways; but, as a whole, what is sown is what is grown: a tomato seed or a seed potato grow into tomato and potato plants. If, then, a human being is conceived then nature will unfold the person who is present from the beginning.

In one crucial respect, however, human beings and plants are different. Although plants can grow wild, as indeed children can come to exist in many different environments, it is God who, on the coming together of human egg and sperm, brings about the embodied existence of the human soul: the child who is both male or female and a being-in-relationship. Just, then, as the embryonic plant is a whole, ready to germinate seed, so God brings about the unique wholeness of the human person by animating the parental ingredients with an embodied human soul; indeed, a human soul cannot but come to exist as the soul of a specific person and, therefore, the instant that a particular human being begins from the contact of sperm and egg then God gives the life of the bodily person.

Tomato Seeds, Plants and Conception

A seed is by definition an embryonic plant[54] which, once it starts to germinate because of soil, water and warmth, begins to show what it is. Tomato seeds grow into tomato plants. When the plant matures, a flower appears and is pollinated which, as we have discussed, brings about the existence of the embryonic plant in the tomato seed. The flesh of the tomato forms around the seed, swells and turns red. We can see, then, that it is natural for the seed of the tomato to be formed within the tomato itself; however, as we have noted, the tomato seed is now fertilized and will grow, all being well, when planted in warm, moist soil.

The woman, however, carries within her the development of her eggs which, in the course of early womanhood, ripen one at a time during the years that it is possible for her to conceive a child; however, the woman's egg is not fertile at this time – rather it is ready to be fertilized. At ovulation, the mother's egg is released and is carried down an internal pathway to meet the coming of the father's sperm, which already indicates that the child, once

---

54 "Seed": https://en.wikipedia.org/wiki/Seed.

conceived, is independent of the mother; not independent in the sense that the child does not need her maternal love, but independent in the sense that the human embryo is a new being who will continue to grow. Just, then, as the mature tomato carries the embryonic plant seed within it, so the mother carries the child within her. Indeed, just as drying the tomato seeds prior to planting them demonstrates the "separateness" of the tomato seed from the parent tomato plant so, once conceived, the mobility and ongoing development of the human embryo, helps us to see that the child is neither wholly from the mother nor wholly from the father – but equally from both and, ordinarily, is orientated to birth, both because of the natural development of the child and the mother's increasing readiness to bring about the delivery of her child. There is clearly no valid implication, then, that bearing a child makes the child a "part" of the mother's body; conception is orientated, from the very beginning, to the birth of the child.

Chapter One: Imaginative Use of Gardening and Plant Life 89

The Early, Hidden Development of both Tomato Seed and Embryonic Child

The "sperm" of the husband and the "egg" of the wife are formed during their ordinary, everyday growth, from embryo to adult. The sperm of the husband is not passive. Just as the pollen, producing the plant sperm, goes down into the ovary of the tomato plant, is an active process, so the man's sperm, once passed to the woman, moves to seek out the woman's "egg". The woman's "egg" is not closed, like a boiled egg, but is rather more like a large ball of wool, which has a host of openings in which a sperm can come to rest in one of them. While, however, the sperm of the husband is sown in the body of his wife, the wife's "egg" is released internally and, if fertilized, then a child is conceived and will be naturally transported to the mother's womb to implant and to continue to grow.

The husband's sperm is very small and travels, as it were, to meet his wife's "egg" which is comparatively large and, not only does the egg contribute to the child's genetic inheritance but, in reality, from fertilization onwards, the egg is substantially the body of the child from which all development subsequently proceeds. On the one hand the husband's sperm, although like a cotton thread in comparison to his wife's egg, which is like a "ball of wool", is nevertheless as dynamic as the egg is inert; and, while this is regarded as controversial, it is based on the evidence of the egg's inactive mitochondria, the deterioration if the egg is not fertilized and the multitude of changes which literally spring from the egg's reception of the sperm[55].

---

[55] Cf. Even a source with which I disagree on so many points says: 'Following fertilisation and activation by the sperm ...' on p. 595 of Anastasiadou and Joep Geraedts, *et al*, "The interface between assisted reproductive technologies and genetics: technical, social, ethical and legal issues". Cf. also Etheredge, *Scripture: A Unique Word*, "Chapter 12: Life from life: A Reflection on the Moment a Person Comes to Exist", pp. 320-322, Newcastle upon Tyne: Cambridge Scholars Publishing, 2014; and indeed, two articles were published: "A Person from the first instant of Fertilization", *Catholic Medical Quarterly* (August 2010, Vol. 60, No. 3, pp. 12-26); Part II of II: "A Person from the first instant of Fertilization", *Catholic Medical Quarterly*, (November 2010, Vol. 60, No. 4, pp. 20-26). Cf. also "The Mysterious Instant of Conception", *National Catholic Bioethical Quarterly* of America, Vol. 12, Autumn 2012, No. 3, pp. 421-430. More information about published work on human conception at the end of the present book.

Chapter One: Imaginative Use of Gardening and Plant Life 91

But these inanimate comparisons do not do justice to the organic structure of sperm and egg. The woman's "egg" has pores which, once they enclose a specific sperm from the husband, naturally close and encase it; hence, the woman's egg is both a receptacle of the husband's sperm but also the foundational "flesh" of the child's body. On the other hand, the animation of the sperm is "transferred" to the egg and thus the human embryo comes into existence. Just as the closing of the pores of the egg is a sign of fertilization, so this outward sign of fertilization signifies the first instant of the human embryo's existence: the first instant of the human embryos existence being the moment of God's ensouling action.

If only it were a matter of evidence. In the emergency of transferring a once frozen human embryo, which is now thawed, it has been possible to see that as soon as the child, Hannah, was thawed, her cell multiplication began again[56]; indeed, what would be the point of arresting the growth of the human embryo if she was not alive and growing in the first place? In other words, the whole nature of fertilizing a human egg outside of the body testifies to the humanity of the child conceived and cries out for

---

[56] Strege, *A Snowflake named Hannah*, p. 105.

the justice of the child's completing human development in the womb of a woman – just as there was an answering "yes" from her adopting parents, John and Marlene Strege. Moreover, the very independent existence of the embryonic child, sitting on a glass dish or frozen in liquid nitrogen, testifies to the reality of a child's existence being other than that of being a "part" of the woman.

The roots of the tomato plant, similarly, are not normally visible except when the plant is being moved or it is in a glass or transparent plastic container. The plant, then, is capable of being transplanted and, therefore, if the little plant begins indoors, it can both be put in a bigger plant pot as it grows and, eventually, it can be planted in a bucket or the ground, ready for when it will mature and start to produce tomatoes. The human embryo, as we have seen, has been known to be "transplanted" if he or she "implants" outside the womb; however, as we have also noted, this has not happened very often and there is, currently, a project underway to help both mother and child when this occurs[57].

---

[57] Margaret Peppiatt, July 22nd, 2022, "Treating ectopic pregnancies is not abortion. But researchers are still looking for a way to save both mother and child":

The fact that the child lives in the womb of the mother, initially, is not unlike the seed of the tomato plant being first sown in a plant pot until such time as it is sufficiently well started to be planted out in the ground. The main point of the comparison being that the soil and the plant are different. Just as the tomato seed is planted in the soil and benefits from it, so the child is conceived in the mother and benefits, immeasurably, from her maternal care. As the tomato plant and the soil are separate, so are mother and child; although, in the case of father, mother and child, their relationship is forever, while in the case of plant and soil, their relationship lasts as long as the plant lives.

Conception and Growing Potatoes

Potatoes, however, while they can be called "seed potatoes" are otherwise known as "tubers" from a Latin word

---

https://www.americamagazine.org/politics-society/2022/07/ 22/embryo-transfer-research-ectopic-pregnancies-243400; in the same article there was a second, successful transfer, in 1990, by Dr. Landrum B. Shettles, of a 40 day old child into the womb of the mother.

which means edible root[58]. Thus, by contrast with the tomato, which grows behind the flower and hangs off a stalk on the plant above the ground, the potato grows underground. Potatoes come in various sizes and have an "eye" from which the new potato plant will grow; indeed, potato plants are remarkable for the fact that they will grow from the slightest piece of potato provided, that is, it has an "eye". Although the potato can be planted without any obvious growth, it is interesting to watch for the start of these developments and then to plant them; indeed, when multiple eyes have sent shoots out, it is possible to break or cut the potato up, giving each piece a "shooting eye" – but no smaller than a golf-ball. However, I have seen slices of potato, no bigger than a large coin, with a root and stem.

When the flower on the potato plant has come and gone, and it is time to dig it up and see if there are any potatoes in the ground, look out for the little tubes which run from the main plant to the potatoes themselves. Thus, the potatoes grow in the ground at the end of little tubes which come off the main plant. This is as near to an

---

58 "Tuber": https://www.vocabulary.com/dictionary/tuber#:~:text=A%20tuber%20is%20a%20plant,the%20plant%20considered%20a%20tuber.

umbilical cord as I have ever seen in a vegetable; and, in its way, makes the growth of the little potatoes a vivid illustration of what happens with the growth of our own children.

More widely, learning how nature operates enables the parents to recognize the time when a child is more or less likely to be conceived[59]. Thus, while accepting that a child may be conceived unexpectedly, there is nevertheless an opportunity to space the conception and birth of children rather like a gardener gives room to each plant. At the same time, however, a person is different to a plant and cannot be weeded out once sown, as the conception of a human being enters that child into the rights of the human race: a human being, on conception, has a right to completing human development: to that process whereby the person present from conception becomes visible. Just as with a potato that has many sprouting "eyes" it can be split into many potato plants, each with an eye, so it is possible that twins and sometimes more children can come

---

[59] See the forthcoming *Human Nature: Moral Norm* on the extensively good work of Natural Family Planning: https://enroutebooksandmedia.com/humannature/.

from the separation of a single human embryo as he or she develops.

The Integrity of Human Being

Tomato seeds grow into tomato plants and tomato plants grow tomatoes and tomatoes have seeds which can be dried and sown and grown into tomato plants. While it is true that there is a natural integrity to plant, animal and human identity, it is also true that it has long been the custom to cultivate plants and animals in ways that alter them. In the case of human beings, however, because this involves the life of a person, who has a right to being loved as who he or she is, it is not right to treat the human person *as if he or she is a product to be manipulated.* A justified intervention for the sake of the health of the child is one thing; but an "experimental" intervention for the sake of the experimenter is totally different.

On the one hand, then, it is quite clear from human growth that we consume a wide variety of ingredients, whether from animals or plants and that these nutrients, ordinarily, contribute to healthy growth. On the other hand, introducing genetic or reproductive tissue from an animal, or even a plant, either into the moment of conception itself or into the early stages of human development, raises extremely serious questions of what is, unless known to be otherwise, a human being with a mixed human and animal inheritance. The claim that we are all animals, so there is no difference between "us" and "them", and so we can mix genetic or reproductive elements regardless of species, ignores what common experience has shown: that many attempts at cross species reproduction are infertile. The number of factors influencing the compatibility of species already makes this mixing of genetic ingredients perilous from the beginning.

Furthermore, however, as the human embryo develops within the mother, it is now more clearly recognized that there is a genetic dialogue, called "epigenetics", which takes place between the child and the mother whereby each subtly influences the other[60]. In a word, then, it is a gross simplification to describe us all as "animals", as if

---

[60] Cf. Francis Etheredge, *Conception: An Icon of the Beginning*, Chapter 5, Part II, prepared by two experts, the first of whom once assisted the *Pontifical Council for Life*, entitled: "The Biological Status of the early Human Embryo: When does Human Life Begin? A Paper by Professors Justo Aznar and Julio Tudela": https://enroutebooksandmedia.com/conception/.

there is no differentiation between species which makes certain combinations "problematic"; and, indeed, so subtle are many of these differentiating features that even a slight change may have a profound effect on human development. Thus it is no justification to say that as a child can be aborted, we can do what we like with it; as, already, an unwarranted decision has been made about our relationship to that child. Each of us is a gift and no one has the right to take that gift from another; indeed, even adults who have committed crimes, it is argued, can be securely detained in the hope of reform, rather than being executed.

In general, then, irrespective of whether a person may think that what he or she is doing does not involve a human life, there is sufficient uncertainty to denote an absolute prohibition on human-animal species experiments. In the end the Creator may well accept that a creature is human, even if it is in fact ambiguously so, precisely because God sees the person that has come to exist, however imperfectly, is still a human person. Just, then, as the Creator ensouls a child conceived out of wedlock, in a test-tube or because of rape, out of love of that child – precisely because the child is not responsible for the imperfect conditions of his or her coming to exist, so the Creator recognizes that what constitutes human being, however imperfectly, calls forth the power of God to ensoul him or her on coming to exist. At the same time, however, the injustice the child suffers on being brought into existence in a certain way is not justified by the good that God brings out of it.

Human Sorrow and Plant Loss

It is true, however, that just as there are various plant diseases and problems, so there are problems with the first stages of human life; indeed, conception does not always occur and, even when it does, the child may die. One of the greatest consolations, when it comes to the loss of a child, is simply the nature of enduring life. On the one hand, from philosophy we can conclude to the enduring nature of personal life because the existence of the human soul transcends the physicality of being embodied; and, therefore, even if the child was lost in a very early miscarriage, the very nature of a personal life endures with the enduring nature of the soul. On the other hand, philosophy is complemented by the Christian Faith, which expresses the belief in eternal life or, conversely, if this is rejected, eternal damnation.

Similarly, in the case of what is thought to be a pregnancy, there is no child but an amorphous mass of cells. In other words, something can so radically alter what happens after the contact between sperm and egg that there is no transmission of life but, rather, a growth which is no more like a child than the sticky mess of a spoilt, germinating bean, is like a bean plant. The technical name for an amorphous mass, following the contact of sperm and egg, is that it is a hydatidiform mole: a name which comes from two words, 'the Greek word *hydatis* meaning a drop of water, and the Latin word *moles* meaning a mass'[61].

In other cases, as with a reasonably robust seedling that is put out too soon and the frost shrivels its leaves and it dies, the child is actually conceived and begun to develop but for whatever reason, sadly, the child is lost through a miscarriage: passing too early from the mother's womb. However, no matter how "precious" our plants become to

---

[61] There can also be a partial hydatidiform mole, too, which usually ends with the embryo being miscarried and lost:

https://link.springer.com/chapter/10.1007/978-3-642-84385-3_18; indeed, it is very difficult to tell if there is a true child present but, nevertheless, if there is a child then there is the grief because of the loss of that child.

us, there is an incomparable dignity and grandeur to the very person God brings to exist – precisely because his or her existence is totally beyond us and speaks of the very mystery of being made in the image and likeness of God.

In general, then, it is possible to open up the whole subject of life, and human life, through growing vegetables and to help people understand the simple truth that from a tomato seed comes a tomato plant, and from a potato comes a potato plant; and, therefore, from the coming together of man and woman, through the embrace of marriage or the marital embrace, comes the fruit of children. But even if, for whatever reason, a child comes to exist for any other reason, God loves the child as the good gift of Himself to us and, therefore, we are called to love all who come to exist too; and, as such, to love is to do good and the first good is to help the life that existed manifest the person who was conceived from the first instant of fertilization.

Chapter Two

A Unitary Beginning of One or Many

Historically, there are both a variety of answers to the question of when we begin and a variety of disciplines which ask and answer this question. To some extent, these attempts are a foundation for our current discussion because, one way or another, either they are pointing towards where we are now going or they need to be discarded, as interesting or as useful as they once were. Traversing this ground, as it were, sharpens our perception about what is involved, both methodologically and actually, in answering this ever-present question of when each of us began.

Two Views: Immediate or Delayed Animation

On the one hand there is the view that the beginning of the human person is after the very first instant of beginning; and, therefore, that the person is present from any time after, or from, the fusion of the two pronuclei, one of which came from the male sperm and the other from the female egg. In other words, according to this view, the person is not present from the very first instant of fertilization.

On the other hand, the embryological evidence, philosophical reasoning and theological thought influenced by *the word of God*, indicate a simple beginning which is the very first instant of active contact between the sperm and the egg. In a word, conception means, a beginning, and just as we recognize beginnings in all walks of life, so we recognize that the word has a specific sense. Thus, there are activities that lead to a beginning but are different to it; and, therefore, after is not the same as before and is different from it. Training for a race is different to running the race; walking to the start and waiting to run is different to hearing "go" and setting off; mixing ingredients, as we have discussed, is prior to, but not the same as the irreversible baking of a cake. A child, then, comes to exist both because of the union of sperm and egg and the hidden action of God which goes beyond and completes what is visible. Just as a variety of methods tends to a single goal, so the different disciplines tend to a single grasp of the truth that each one of us has a beginning.

Each discipline, whether it is embryology, philosophy or theology, has asked the question of when a human life begins and none of them, it seemed, had a uniform answer. This shows, in its own way, that the questions we ask are influenced by a whole range of factors and, at the same time, that there may be a particular difficulty in answering the question itself. Inevitably, then, there is a variety of answers which, down the years, have tended to recur, albeit inconclusively. Nevertheless, reviewing these answers is an age old method of both gathering what is good and answering the questions that they raise with what has subsequently come to light.

The possibility of a Single Answer to When Did I Begin?

What makes the possibility of a converging answer possible from these different disciplines? There are two, if not three answers to this question. Firstly, each of them is interested, as it were, in the question of beginning; and, therefore, there is a common "object" of any inquiry which has arisen concerning the beginning of the human person. In other words, the beginning of the human person stands as the "reality" to be understood, both now and at any time; and, should there be any doubt about it, each of us is a living witness to having had a beginning and to this being a real question: What defines the moment of our beginning? Secondly, the existence of a variety of answers to the moment each of us begins does not necessarily suggest that there is no answer as that the question of "when" needs to take account of each discipline's tendency to answer; for, if the object of our inquiry is a real object, then each discipline is a part of the whole response of the human race to this question. On this basis it is clear that truth will not contradict truth; but, according to whether or not it is embryology, philosophy or theology, the emphasis or preoccupation may be different. Thirdly, taking the unitary view that creation is an act of God, however that is to be understood, there is the coherence of these disciplines in view of the fact that each one of them is answering the question according to the truth that has come to be known or revealed.

Embryology: What is One Organism?

What makes an object one? Numerically, it can be argued, one is an adjective which describes a single object; and, therefore, let us consider a single, embryonic plant seed. I am thinking, then, of the courgette seed which, although we don't ordinarily see every step of its germination or growth in the soul, is yet a good example of a very visible seedcase. As the temperature rises, in a moist soil, and the seed begins to germinate and to grow, it pushes down its roots into the soil and its leaves slide, as it were, out of the seedcase. Thus, there is a very visible plant and seedcase; indeed, as the leaves grow, so the seedcase sits on the edge of the leaf like a hat. In other words, the seedcase is clearly very different to what grew out of it.

Now it is not possible to speak of this being like the mother's womb because the womb is integral to the mother; however, just as the plant leaves the seedcase, so the child is born from the mother. On the one hand, the seedcase does not appear to have been alive at all except when it originally grew and, as such, was not alive of itself but was a kind of protective sheath of what was dormant and, in due course, ready to grow. Thus, the drying out of the seed case is not necessarily the drying out of the seed. Nevertheless, come germination, the courgette seedcase is simply discarded. One seed, then, is both case *and* embryonic plant; and, as the embryonic plant grows so it is clearly separate from the seedcase. In this instance one seedcase encases one embryonic plant. On the other hand, the mother of the child is very much alive and active in becoming pregnant, bearing and delivering the child into the world; however, just as the seed-case encloses the embryonic seed, which it leaves as it grows, so there is an intra-uterine environment within which the child develops, known as the amniotic sac, containing fluid, umbilical cord and all that is necessary for the child's growth in the womb. Clearly, just as with the seedcase, the growth of the child towards birth entails reaching the limits of the womb and, in due course, leads to the delivery of the child both from the womb and the encasing amniotic sac which surrounded the boy or girl.

The courgette plant, then, is both root and shoot, discarding what was helpful but now redundant. A whole plant includes, then, whatever is integral to what makes it one plant: root, shoot and, in due course, both flowering and being fruit bearing. When it comes to a bunch of leek seeds which have been planted in a small pot, they eventually out-grow their container. Shaking the soil off the leek plants shows that their roots are a bit tangled together; however, as they are gently pulled apart it is quite clear that they are a group of separate plants, each of which has now been planted and is growing independently of the others. Just like children conceived from separate eggs, the twins or triplets develop independently of each other as indeed the birth of a girl and a boy, or two girls and a boy make clear.

All in all, just as each type of being, whether plant or human, has many characteristics, all of which express the whole, so the whole is greater than the parts and is integral to it. A human embryo is one human person and, if there is twinning or triplets, then there are two and then three children from the same or multiple eggs. In other words, a little examination of reality shows that there is no confusion about the fact that one divided by another is two: one plus the original one; and were there to be any possible doubt, then conjoined twins show us the reality of two people from a single, albeit incompletely divided, human embryo. Conversely, if the conjoined twins have arisen from two distinct embryos, known as non-identical twins, partially recombining, then the point remains that two bodies, however combined, are the bodies of two people. In other words, even if at times it is difficult to determine the precise beginning or end of the body of one person and the body of another, it is nevertheless clear that a body expresses the temporal reality of the person in an unrepeatable way: one body expresses one soul and the whole is the human person.

Reverting to What Is Original

Returning more explicitly to the theme of unity there is, surprisingly, what happens to the heart of a cabbage. Take a cabbage, one already cut from the ground and packaged, put in a supermarket and taken home, eventually, sliced up and prepared for dinner – and you take the heart of that cabbage, normally bitter and hard and discarded and you plant it! What do you expect to happen? That the cabbage starts to discolour and to rot, maybe it will get a bit eaten as a meal for slugs and snails or simply, unobtrusively, shrivel away. But would you expect it to grow into a cabbage plant? But that is what happened – not once but to three or four cabbage hearts planted in the soil with a bit of compost and watered! In other words, and this is the point, there is not only a residual life, calling into question the actual "death" of the plant but there is a definite cell development which is not in the least random but is in fact a reversion to the whole process natural to becoming a cabbage plant – not growing cabbage-like but actually growing into the cabbage plant from which flowers and seeds come!

There is, then, a "residual life" in what appears to be a "dead" cabbage: decapitated, rootless, left on a shelf and cut up for cooking; but, in and amongst the cells of that discarded, bitter heart, are what are still capable of growing, given the ground into which it is planted and the normal conditions of growth. We notice this too with transplanting lettuce, leeks or other plants, that if they are forgotten and left out overnight on the ground they do not just wither and die in a moment but they remain alive and ready to grow for a definite period of time Thus it is possible to speak of a real presence of life that persists even when the conditions for its expression are temporarily withdrawn or problematic. Similarly, in the case of those of us who have died through heart failure or for whatever reason, there is a possibility of being revived; and, as we know with resuscitation, the person can be revived and live and even live a long time after being revived. Thus, the whole concept of dying needs to take account of the residual nature of life and not anticipate death impatiently or even for reasons of organ stripping[62].

---

[62] The concept of "brain death" has this disadvantage that if the major organs are functioning, such as the heart and lungs, there is the presence of life and therefore of the human life of the person. In other words, the person is not dead; cf. *Reaching*

Furthermore, the conception of any child, under whatever circumstances, is going to entail a life that persists until it is definitively dead; and, therefore, life is integral to what is conceived and begins to unfold and manifest its personal nature in the course of his or her development. What is more, the life that persists is the life of the whole person and not just some functioning of a part of the whole; and, therefore, where there is the life of a person there is the whole life of that person.

In a dramatic illustration of an embryologist's discovery, contribution to his conversion to the realization of the beginning of human personhood, he says:

> 'Despite having some cells that are abnormal, an early embryo has the capacity to "self-correct." It does this by selectively pushing the abnormal cells out and replicating the normal cells. In fact, it appears that the abnormal cells become part of the placenta, leaving the normal cells to become the fetus. This is a remarkable

---

for the Resurrection: A Pastoral Bioethics: https://enroutebooksandmedia.com/reachingfortheresurrection/.

biological process that is somehow "programmed" into these primitive cells'[63].

A Difference of Words: Embryo and Child?

Without going into the details of the case, what is at issue here is the following claim: 'In its justification the [European] Court claimed that the notion of a *child* belonged to a different category than the notion of *embryo*, and that these two terms could not be viewed as interchangeable'[64].

---

[63] Dr. *Craig Turczynski is a Reproductive Physiologist, Certified Teacher of the Billings Ovulation Method®, Director of Strategy and Scientific Affairs for BOMA-USA, and currently serves on the board of advisors for Sacred Heart Guardians and Shelter:*"Abnormal Embryos and Human Life: An Embryologist's Post-Conversion Reflection": https://sacredheartguardians.org/abnormal-embryos-and-human-life/. *More on his testimony in the preamble to Part IV of this book: Conversion to Reality.*

[64] Appendix: Case Illustrations, p. 119 of Alichniewicz and Monika Michalowska "Medicine of the Beginning of Life." Etc. However, the further point that the European Court is relevant

In the first place, let us note that even the Warnock Report recognized that there is a seamless process of development: 'there is no particular part of the developmental process that is more important than another; all are part of a continuous process'[65].

Secondly, while the notion of human "embryo" and "child" are clearly different – they refer to different stages of the same continuous process. Referring back to the discussion on the use of imagery, would we distinguish the diver, with all his equipment that keeps him alive

---

but beyond the scope of this discussion: 'Additionally, the willingness expressed by the government to defend values stood at odds with the possibility to carry out an abortion in the case of a diseased foetus, which as a procedure involved negative consequences also for the woman': European Court of Human Rights, Case of Costa and Pavan v. Italy, Judgement, 28 August 2012 (Final 11/02/2013), http://hudoc.echr.coe.int/eng?i=001-112993 (ac-cessed October 15, 2019) – as per the original author's document.

[65] Department of Health and Social Security [UK], *Report of the Committee of Inquiry into Human Fertilisation and Embryology* (London: Her Majesty's Stationery Office, July 1984), para. 11.19.

underwater, with the person in the suit? In other words, it is perfectly valid to recognize the difference between the human embryo, which is like the diver and his equipment for staying alive, and the person who, having stepped out of his suit, is clearly separate from it.

The problem, then, is an inadequate understanding of the relationship of embryonic development to the manifestation of the human person; and, therefore, distinguishing "embryo" and "child" to the point where they are not "interchangeable" is to deny the relationship between the diver and all his equipment and the person who steps out of the suit!

Part III

Literary Truths and the Literal Truth

COMPRISING CHAPTERS THREE AND FOUR

What is an analogy? An analogy is a partial likeness between two different things in order to help us to understand what is less visible but still true. The point of this brief discussion is simply to indicate how an author can use imagery to indicate partially hidden mysteries; but, in this use of imagery, there is a distinct sense of the author drawing on what exists but not being restricted to a literal use of it.

When God made man and woman in the first book of the Hebrew and Christian Bible, we can see that the opening chapter gives us a varied account of how He brought to exist the whole of creation; but, in that first chapter, the author does not explain how God created man, male and female, although we can see that man and woman are made for "relationship" just as God is a mystery of relationship[66]: 'Let us make man in our image, after our

---

[66] For a discussion of the three names of God in the first two chapters of Genesis in the light of the Hebrew text, go to Chapter 8 of Francis Etheredge's *Scripture: A Unique Word*: https://www.cambridgescholars.com/product/978-1-4438-6044-4.

likeness …. So God created man in his own image, in the image of God he created him; male and female he created them' (Gn 1: 26-27). In the second chapter, then, the author says: 'the Lord God formed man of dust from the ground, and breathed into his nostrils the breath of life; and man became a living being' (Gn 2: 7). Thus, we see that the creation of man is distinguished from the creation of the animals in that there is a personal action of God which brings the man to exist – both from the earth and because of the breath of God. In other words, God makes man a unified being of both matter and spirit. The emphasis, then, in the making of woman from the man, is that she is "life from life" and, therefore, a sign of "relationship": the relationships hidden in God but indicated in the analogy of man and woman's being made in a unique way. Just as the Son of God comes forth from the Father and the Spirit proceeds from both of Them, so man is made from the earth and the woman is "from" and "through" the man.

In the case, then, of the breath of God, we do not understand breath as indicating the bodily presence of God – but the power of God to give, personally, a personal life to man: a life that is intimately and mysteriously shared between the first man and the first woman – each being equally a person. Similarly, when it comes to thinking of the 'dust' from which man is made, it is clear that on the one hand man is made of materials common to the whole of creation – but which are lifeless without the gift of life given by God; but, on the other hand, it is also clear that when God 'formed man' he made this dust capable of bearing the spirit that gave life to him. As another author says later, when it is better understood that there is both a beginning and a development of that beginning: 'Thy eyes beheld my unformed substance' (Psalm 139: 16) and 'For thou didst form my inward parts' (Ps 139: 13) – as the author moves from what is developing to that from which "I" developed; literally, in the Hebrew, "*golmi*", my unfinished vessel![67]

In contrast, then, to this literary use of imagery to communicate a variety of truths, is what constitutes the literal use of understanding one thing, not as like another, but as some kind of real version of the other but in a very different context. In what follows we will examine a line of thought which took the view of a three-stage stepped beginning to human personhood, the first step of which

---

[67] Cf. Chapter 11 of Etheredge, *Scripture: A Unique Word*, 2014.

proposed the existence, literally, of a plant type of being to the beginning of human life.

Chapter Three

Passing Through the Past to the Present

The idea that there is not an immediate beginning to being a human person but that there is a process of becoming human has both a very ancient lineage and very modern currency. Therefore it is necessary to examine this kind of account and to show the difference between an author literally understanding the beginning of human being to be that of a plant, or at least not fully human, but becoming so in stages.

From the "Literal" Use of an Image to the Truth of Embryology

Going back, then, to our discussion on analogy, it is helpful because, as we have seen, there is a partial likeness between two different subjects, the more familiar one helping us to understand the less familiar one; for example, the lifeless 'dust' being formed in such a way that it can receive the breath of life, where breath is commonly understood as a sign of being alive. The contrast of being alive with dust is more about the wholeness of life coming from God than, literally, we are made from the dust of the ground; however, as a figure of what we have in common with the matter of the universe, "dust" serves very well. We now come, then, to a literal use of a comparison between plant and human life. But first a little context to help us to understand where our philosopher is coming from.

Aristotle (384-322, BC) was an early philosopher who researched and wrote on a wide variety of subjects. In general, he held that God was a "First Mover" of what exists and, unlike the biblical Creator, the "First Mover" did not bring matter to exist but, rather, brought about changes in what did from all eternity already exist. Aristotle held that matter had always existed and so what changed matter, was the "form" that it received; and, therefore, the "First Mover" was the origin of all change in matter. On the one hand, a piece of wood exists but according to what the carpenter does, that wood is a chair leg, a table or a shelf. In other words, there is what exists and then there is what changes what exists from one definite object to another – from a piece of wood from a tree to a chair leg. In the case, then, of a plant there is that which gives life to inanimate matter, matter that is non-living, and that is called, in this case, the plant soul: that which makes matter not just to be a particular plant but to be a living plant. Similarly, an animal soul makes non-living matter to be a living animal, which moves, grows and reproduces. Lastly, a human or rational soul makes it possible for non-living matter to be a living human being. On the other hand, to understand the relationship between "matter" and "form" it is necessary to think of "form" as radically changing matter – not just shaping the wood as a carpenter would but bringing about the very change in matter that makes it wood in the first place, or a plant, or an animal or a human being. In other words, "form" intimately determines what matter is: a plant form or soul makes the matter to be a living plant.

Stage One: Taking the Comparison with a Plant to be Literally True

Aristotle, then, considered the first instance of human development to be more than comparable to the growth of a plant. He thought this, mistakenly as we now know, because he thought that the initial condition of human growth was a kind of thickening of blood; indeed, that the male semen brought about a change in the blood which completed the first stage of development. In other words, given the simplicity of what he understood to be the starting point of embryonic development, he could not envisage a more developed or complicated being arising out of the clotting and forming, as it were, of blood into a first being. Thus, his whole understanding of there being a plant type of initial human development is predicated on his belief that little more than liquid ingredients were needed to begin human development. Now, then, it is perfectly plausible that the sperm introduced a kind of "order" or "form" to this "matter" such that it transformed it into a plant type of existence. For, on the basis of blood being the first "matter" with which human being begins, it was clearly impossible for it to be "instantly" developed into what would be capable of receiving an intelligent, rational soul.

Aristotle thought, then, that the first step of human development is literally like the growth of a plant. Thus, he thought that a plant soul inwardly shaped human matter to be a kind of plant; but, as we now know, Aristotle's understanding of a human being's beginning as a plant, actually works very well as an analogy but is no longer literally true. Although the human embryo is like a germinated seed, taking in nutrients, im-*planting* in his or her mother's womb and, through multiple cell-divisions, increases in size and complexity with a more and more

Chapter Three: Passing Through the Past to the Present

recognizable differentiation of parts – it is the growth of what is from the beginning a human embryo and not a plant. In other words, while Aristotle could not appreciate how wonderfully and intricately ordered the human embryo is, and therefore he resorted to a simple explanation of its development, the sophistication of the human egg is such that fertilization initiates immediate and ongoing changes.

Simultaneously, as we now know, some cells develop into the placenta, the organ between the mother and the baby, which moderates and facilitates the transfer of nutrients to the baby and the excretion of waste products from the baby through the umbilical cord to the mother's blood supply and, ultimately, to be excreted by the mother. What has been understood, then, is that the baby begins his or her life as, literally, being planted in the womb of the mother. This is not a diminution of the dignity of motherhood; it is, rather, a true recognition of the fact that the child is not a part of the mother but the mother is "bearing" the child. In other words, the ancient philosopher understood very well that the child has a "beginning" and that that beginning entails concrete steps which manifest and unfold the reality of his or her independent existence.

Stage Two: Movement, Sensation and the Rearing of Young

The philosopher's second stage of human development is when there is movement and sensation and, while a lot later, the capacity to bear and rear young.

The philosopher, to explain the change from the planting of a human embryo or germinating seed to being capable of movement, suggests that just as there was a plant soul that changed matter into a living plant so there is an animal soul that changes matter into a living animal. We must remember, too, that the word animal comes from an expression "to animate": to make to move. In other words the philosopher was suggesting that just as human development goes through another stage of development, which now entails being no longer planted in the womb but being able to move in the womb, then he or she is given another kind of soul: an animal soul which makes movement possible.

By way of a little explanation, we can see that a plant responds to temperature and that when it is cold a flower will often close up and that when it is warm that same flower will open up; and, more generally a plant, without getting up and walking, will nevertheless lean in the direction of the sunlight and turn, slightly, to follow it. In other words, as simple as it is, the plant is capable of limited movement and thus is ready to receive an animal type of soul that provides, therefore, for the further development of movement. An animal type of soul, enabling movement, makes possible all the characteristics that belong to an animal, such as sensation, digestion, the power to be able to reproduce and the limited horizon of understanding that belongs to making rudimentary dwellings, finding food and rearing young.

In reality, however, we now know that after the implantation of the embryonic child, the developing embryo is rapidly generating the beginnings of its mature form of head, brain, spinal cord, heart and heartbeat, limbs and the whole interiority of organs and their coordinating functions. In other words, movement and sensation are not because of the imposition of a second form on "matter" but because the human embryo, already moving through a tremendous amount of cellular development, is developing according to the inbuilt human pattern, developing all that is entailed in having limbs and, what is required to move those limbs.

Furthermore, too abstract an account of the developing child, or too great an emphasis on intellectual powers, overlooks almost entirely that conception is about relationship; and, therefore, the more "visible" becomes the presence of the child the more explicit becomes the relationship. Loss, therefore, as in a miscarriage, is clearly a fruit of the relationship that has become more and more explicit; relationship being there from the beginning in the very nature of conception, so that even the early loss of the child is expressed, therefore, in terms of the grief of the mother, father and other relatives.

Stage Three: Rational Ensoulment

We are now, therefore, in a better position to understand the ancient philosophical idea of rational ensoulment: that Aristotle, understandably, thought that there needs to be a sufficient degree of development so that the creature could be ensouled as a thinking human being; indeed, a rational human being is more than just a thinking type of human being, for reason structures the whole being in such a way that we are now talking about the whole human being. Thus, as the child matures and readies for birth, there begin to be the signs of distress, contentment, and the rudiments of play.

By now, however, what we are actually witnessing is the interior structure of the human person maturing, physiologically and, therefore, psychologically, so that there is not an additional action of a different kind of soul so much as what is there is increasingly manifest in all his or her rudimentary humanity. By comparison, then, a computer program may be layered in such a way that to begin with there are simple operations that, once mastered, lead to the unlocking of more and more sophisticated capabilities. In other words, it is not that there is present a new kind of soul as that physiological development allows the showing forth of the more subtle, expressive and intellectual potentialities which are now able to start to show themselves. Indeed, just as a child masters walking and running, so he or she begins to demonstrate the existence of balance, control of a ball or even the dexterity that is a part of being a musician.

While we have seen that this ancient idea that the human person began in stages; and, while it had some credibility in the past, what we now see is that it was a "theory of gaps": an attempt to explain human development through a comparison with a three-step ladder of being. The problem that this ancient idea attempted to solve was how to go from a simple beginning to a stage of development that would make it possible for God to bring the whole human person to exist, one in soul and body. In other words, it was thought that the very beginning of embryological development was too simple to be the moment in which God could create the human soul.

A Concluding Reflection – Towards Understanding Human Ensoulment

Using what was perhaps more literally understood by the ancient philosopher now, as an analogy, helps us to see how to understand a whole plant, the independent conception and growth of the child and the confirmation, of birth itself, that a child is no more a "a part" of the woman's body than a plant is a part of the soil in the ground. In other words, just as a plant in the soil can germinate, grow, be transplanted and ultimately uprooted if necessary, so it is obvious that the child is likewise "implanted" in the womb of the mother, develops and is delivered at birth. Just as the unfertilized egg is from the mother and constitutes, in its own way, the basis of the bodily being of the human embryo on being impregnated by the father's fertilizing sperm, so the lining of the womb is so ordered to the implanting of the embryo and, therefore, the dialogue of growth, such that the developing embryo can draw what he or she needs to grow from the mother. Naturally, then, the diet and health of the mother are equivalent to providing an enriched soil for healthy plants. What is more, however, the presence of the developing embryo in the mother is also expressive of the centrality of "relationship" to the formation of the humanity of the human being.

In contrast, however, to the progressive replacement of a plant, animal and rational soul, it is now abundantly clear that there is a seamless development from conception to the full manifestation of the presence of the person. Furthermore, what is clearer to us than to ancient philosophers is that the embryonic human child is far more developed than was hitherto understood and, therefore, given that the human soul is the life of the body, no sooner does the body come to exist than the ensouled, whole person, one in body and soul, is present. In other words, although we cannot see the soul, philosophically, if there is no soul then there is no human body – for a body is not a body if it is not specifically animated by a soul just as there is no such thing as a husband or wife who is not a particular person. This is because the unity of body and soul is so intimate that there is no "generalized body" but only the specific human life, bodily expressed, which is ensouled and shows itself to be personal from the first instant of conception, expressing as it does a unique but totally human genetic inheritance.

The Greatest Natural Transformation: The Unfolding of Conception

Modern human embryology, then, is far more definite about what exists from the first instant of human fertilization: from the beginning[68]. Indeed, right from the first instant of a fertilizing sperm in an open pore of the egg, the calcium ions rush from the sperm and across the newly formed human embryo, closing the embryonic wall. Indeed, everything has a precise "moment" and significance in terms of timing, direction, sequence and function – so that the whole being of the human person undergoes what is probably the greatest natural transformation in nature: the unfolding outward expression of the

---

[68] Cf. For an excellent introduction to human embryology, go to: Chapter 5: Part II of Etheredge, *Conception: An Icon of the Beginning*:

https://enroutebooksandmedia.com/conception/, where you will find an account of early embryology by Professors Justo Aznar and Julio Tudela.

inward presence of the person from conception. Just as whatever is biologically alive, expresses its life electrically, so a current being on is inseparable from what transmits it; indeed, just as yeast introduced into dough, spreads and makes it rise, and is no longer distinguishable from it, so human life once begun is irreversibly human. Conversely, it is clear that the various disabilities are not about being anything less than human but about the challenge of communicating that humanity when its natural expression is impeded, frustrated or, in some measure, is dependent on an adequately helpful technology.

Note that the transformation entailed in a human person coming to exist unfolds an inner nature scarcely comprehensible in its initial and first formed stage of being a human embryonic person until, literally, there is the manifestation of the person present from conception. In other words, what if we could run the development of a beautiful flower backwards until it was the first instant of a germinating seed, would it help us to appreciate that running the development of a human person backwards takes us to the first instant of human conception?

Why cannot a human soul be transmitted through the process of human procreation? The human soul is expressed in the body as a whole; indeed, the soul is the form of the body: the soul is that which determines the body to be Paul or Janet's body. Given the life of a plant and an animal, what they pass on is within the power of each to reproduce: to literally replicate what it is to exist as a plant or an animal. However, in the case of a human being, there is an element of organic life which is transmitted through the activity of the sperm – but the power of human thought transcends the power of an organic transmission of life; and, therefore, there is an act of God which establishes the whole human being at conception – not just as a general kind of human being but the human being of a particular human person, capable of expressing what is beyond the capacity of either plant or animal life to express, namely truth and love.

While, then, there are many ways of putting our comparisons between the beginning of one thing and that of another, there is also the literal truth that there is a beginning to human life to which each one of us is a concrete witness.

CHAPTER FOUR

SCRIPTURE AND THEOLOGY: WORD AND DOGMA

The progress of science, however, has shown that from the first instant of contact between sperm and egg, they form the new entity of an enclosed, walled and dynamically developing human embryo. Thus, it is now clear that the dynamism that begins with the sperm's entering the egg, bringing about an irreversible change that makes present the beginning of embryonic development, is a dynamism that unfolds the uninterrupted presence of the person from conception. Indeed, as the embryological development unfolds, so we see that the child's physiological and psychological development shows itself in his or her movements and responses to both internal changes and to the stimuli of being-in-relationship to others – particularly the mother, the father and the immediate family and friends who are "with" and look forward to meeting the child. The psychological development, embedded and expressed in the unfolding of the anatomical and physiological developments, shows that there is present, from the first instant of fertilization, an identity which goes beyond what biological matter, whether sperm or egg is capable of. This is particularly clear in that the un-combined sperm and egg deteriorate, showing that they are transmitting what is biologically required for human life but which are an insufficient explanation of the fullness of human personhood.

The activities of the child which show forth his or her humanity are varied, ranging from responsiveness to the mother and father, humour, expressing the psychological states indicating hunger and thirst and, little by little, taking up the common inheritance of the language and beginning to express ideas. As the emergence of language acquisition, thought and humour show, the child expresses a being which goes beyond the capabilities of organic matter. Thus, the psychologically personal life of the child expresses the presence of what goes beyond plant and animal life, namely the presence of a human soul; and, in view of the seamless, uninterrupted nature of human development, the moment of the existence of the embodied soul is therefore the first instant of fertilization. What causes the existence of the embodied soul, and therefore that of the whole human being, has to be a cause capable of bringing to exist what did not; and, therefore, just as the powers of the human embryo show the presence of what exceeds the capabilities of matter, so the cause of human personhood exceeds that entailed in what the sperm and the egg can contribute to human existence. What the whole of human identity and development communicates to reason, then, is the presence of a cause that exceeds human existence, namely God, just as human existence exceeds the limits of the organic matter through which it is expressed; and, therefore, the existence of the person reveals an action of God which brings about a uniquely embodied soul from the first instant of human fertilization.

The unfolding of that first instant is one of the most amazing journeys of being: that the inwardly present soul, because of the creating cause of God, determines the outward expression of the whole human person. The scientific evidence is overwhelmingly clear that from the first instant of fertilization there is an autonomously functioning being that is wholly independent of the parents and, at the same time, is in relation to both: to both the father and the mother. Each one of us is, then, an implicit witness to the beginning of a new being: the specific person that each one of us is.

The mother, being the bearer of the child, expresses more completely the origin of the child in the love between the father and the mother; and, in addition, the essential truth that relationship is at the core of human existence and development. Thus, the mother's welcoming receptivity is both an expression of what she contributes and is its own evidence of relationship being the fundamental vehicle of human development. This truth of science, both embryological and psychological, is confirmed abundantly from the very different but complementary account to which we now turn.

In the Scriptural text we will see that there is a very personalistic account of the conception of the whole human person, implicitly understood as a "living being"; and, as such, each one of us is an intimately personal work of an intimately personal God. In other words, we see that the word of God completes, in a way, our understanding of human conception and development by taking us further and further into the mystery that *relationship* is the great and understated mystery evident, but not explained, throughout the natural world and particularly present in the conception and development of human life.

But first we must consider, however briefly, on what basis we can justify our use of the biblical account.

The Word of God and Dogma

It can seem to us that the word of God, the Scripture, is not a truly human word or not really a truly divinely inspired word. We might, for example, argue how is it possible for a human author to write anything that is absolutely true when, as we all know, human beings are so prone to mistakes or to find some subjects so difficult to understand that they remain almost totally impenetrable. On the other hand, how is God going to be able to communicate through a word which is so obviously written through all the problems and sins of the human race to the point that one wonders is it even possible that the writer can be sufficiently knowledgeable or open to divine inspiration that, notwithstanding all the author's defects, he can still write *the truth which, for our salvation, God has confided to the sacred Scripture (Dei Verbum[69], 11)*? So we

---

[69] The documents of the Church are generally known by their first Latin words and the number that follows is the paragraph numbering given by the Church when the document is published; *Dei Verbum* means the word of God, literally, and is a text from the Second Vatican Council explaining the nature

come upon all the challenging questions which surround the use of Scripture in a discussion such as this and, surprisingly, we find an unexpected answer in the dogmas of the Church: that there is a certain understanding, made possible through the help of the Holy Spirit (*Dei Verbum*, 5 and 8) which takes us to what is essential for our salvation. 'One ... of Cardinal Ratzinger's most famous statements that dogma is simply the Church's infallible interpretation and elucidation of Scripture'[70]. Moreover, we find the help we need in an unexpected dogma concerning Mary, the Mother of the Lord; but, before we consider Mary, the second Eve, we will start with a few other voices, beginning with the first Eve.

---

of Scripture and its relationship to the teaching office of the Church and to Tradition.

[70] Dr. Joshua Madden, "Newman, Aquinas, and the Development of Doctrine": https://www.hprweb.com/2021/06/newman-aquinas-and-the-development-of-doctrine/.

A Variety of Witnesses to Human Conception – Beginning with Eve

'Now Adam knew Eve his wife, and she conceived and bore Cain, saying, "I have gotten a man with the help of the Lord"' (Gn 4: 1). This is the earliest, simplest, truest, comprehensive and most enduring account of human conception that we have; and, indeed, it will take millennia to appreciate and grasp, more and more fully, the fullness of truth that is expressed by both Eve and the biblical author. On the one hand the biblical author recognizes that 'Adam knew Eve his wife, and she conceived'. Thus intercourse is an intimate "knowing" between husband and wife and, by implication, involves the wholly personal communication between them which is, at the same time, a moment that arises out of the social reality of their marriage: a wholly reciprocal gift of husband to wife and wife to husband of all that each of them is. On the other hand, the author quotes Eve as knowing and saying "I have gotten a man with the help of the Lord" (Gn 4: 1); and, therefore, Eve has an equally intimate sense of the Lord's help giving her 'a man'. Indeed, so significant does she regard 'the help of the Lord' in conceiving 'a man' that she does not even consider what the narrator tells us, namely, being known by Adam; rather, so conscious is she of the 'help of the Lord' that this is her primary perception: that without the 'help of the Lord' there would be no conception. In other words there is a dramatic sense of the act of God which both manifests the continuation of the Creator's act of creation and its essential contribution in bringing a human being to exist.

Job

He builds on Eve's perception, making it his own when, complaining about his own life Job says that God has taken a stand against 'the work of thy hands' (Job 10: 3). Elaborating on his beginning, Job goes on to say: thy 'hands fashioned and made me' (10: 8) and, as if recalling the second account of creation, Job asks God to remember, as he himself recalls, 'that thou hast made me of clay' (10: 9). Job then takes a further step and, drawing on his own imaginative analogy, he starts to analyze the "moment of his conception"; he compares his origin with the curdling of cheese: 'Didst thou not pour me out like milk and curdle me like cheese' (cf. Job 10: 10). Thus Job has expressed himself in a new way, going further into the possibility of the action of God in the mystery of human conception. In other words, taking the literal image of being poured out 'like milk', suggesting that God has not only poured him out but that there is also a likeness between being poured out and milk, implying some understanding of what is involved in human conception, possibly of the semen being like milk and that this is a stage in a process, after which God then curdles him 'like cheese'.

Without, then, attempting to identify too closely this analogy between milk and cheese and human conception, yet one can recognize the imaginative connection between what we now know and what, to some extent, Job surmised from what he knew happened. What distinguishes this process of Job's, however, in contrast to the philosophical steps of three types of soul, is that Job is Job from the beginning, through to the end. In other words, there is a sense of personal identity that runs through the whole event from when Job first recognizes that he exists because of the action of God: he is the 'work of thy hands' (Job: 10: 3); and, by way of continuing emphasis, each stage is still a stage of Job being made: 'Thy hands fashioned and made me' (10: 8) right up to the possibility of him going to the 'land of gloom and deep darkness' (10: 21). By contrast, then, the early philosophical account, for all its richness, does not address a wholly human identity until the third step of rational ensoulment (following the plant and animal stages of development).

Moreover if, taking the relationship between man and God from Genesis as that we are made in the image and likeness of God (cf. Gn 1: 26), then it follows that Job is addressing God as a "personal being" the whole time as, from start to finish, he addresses God as the author of Job's being: Job is the work of 'thy hands' (Job 10: 3) and, when coming to the depth of his suffering he says: 'Why didst thou bring me forth from the womb?' (Job 10: 18).

David

The author of Psalm 139 similarly, recognizes the action of a personal being who has made him: 'For thou didst form my inward parts, thou didst knit me together in my mother's womb' (Ps 139: 13); and, using a unique Hebrew word which occurs only once in the Bible, *golmi*, David says, 'Thy eyes beheld my unformed substance' (Ps 139: 16). The expression, 'unformed substance', more literally means 'unfinished vessel' and, as such, implies what has been made, which is a container, containing either a soul or the grace of God or both, that it is unfinished and therefore developing, and that God is making "my" unfinished vessel as David also says: 'for thou didst form my inward parts' (Ps. 139: 13). In addition, however, David also uses the imagery of a book which is, after all, a personal work of an author; and, therefore, both he and Job, while not giving exactly technical account of their conception are nevertheless indicating sufficiently clearly that there is a beginning and a process of development; and, in their own way, therefore, they are very reasonable accounts of human conception within the limitations of the times in which they were written and the language that was available to them. Although, having said that, David's word, *golmi*, seems to have been specially coined for the task since, as I have said, there is only one instance of its use in the whole Hebrew Bible; indeed, a unique word for a unique event of the action of God and the beginning of a human life.

The Martyred Mother of Seven Martyred Sons

Finally, and somewhat surprisingly, the most explicitly scientific account of human conception and development is from a Martyred Mother of several martyred sons; she says:

'I do not know how you came into being in my womb. It was not I that gave you life and breath, nor I who set in order the elements within each of you. Therefore the Creator of the world, who shaped the beginning of man and devised the origin of all things, will in his mercy give life and breath back to you again, since you now forget yourselves for the sake of his laws' (2 Macc 7: 22-23).

Furthermore, by way of complementing this more speculative account of human conception, referring to 'life and breath' and the fact that the 'elements' were 'set in order ... within each' of her sons, this mother goes on to speak clearly of the process through which one of her sons came to exist: 'I carried you nine months in my womb, and nursed you for three years, and have reared you and brought you up to this point in your life' (2 Macc 7: 27). Calling into the argument the witness of what God made out of 'what did not exist' she went on to say that 'Thus also mankind comes into being' (2 Macc 7: 28); and, therefore, that if God can make all that exists out of nothing so He can return her sons to her: "Accept death, so that in God's mercy I may get you back again with your brothers" (2 Macc 7: 29). This Mother's reflection on the coming to be of her sons is all the more remarkable as she witnesses their martyrdom before, finally, being martyred herself for refusing to reject the binding laws of God. This Mother's testimony is a wonderful combination of faith and reason; faith in the power of the Creator which is, in addition, married to a 'woman's reasoning' (2 Macc 7: 21); and, in the context of her whole testimony, this 'woman's reasoning' is both scientific and experiential, both observant and penetrating in her analysis.

Mary: The Dogma of the Immaculate Conception and Human Conception

We now come to the final point of this examination of the Scriptural evidence for the beginning of each human life, excluding the creation of Adam and Eve. What we find is that the Church has expressed herself in the dogma of the *Immaculate Conception* of Mary, the Mother of our Lord Jesus Christ. A dogma, of which there are other examples, such as the Assumption of the Blessed Virgin Mary into heaven, one in body and soul, on her death, is an expression of the Church's certainty that the doctrine proclaimed is not just true but of a truth so necessary to the whole of salvation that we are called to recognize it as integral to our salvation; and, therefore, to disbelieve or to reject a dogma seriously imperils the wholeness of our faith in what God has done to bring about our salvation in Jesus Christ.

Thus the *Immaculate Conception* of Mary is that she was conceived full of grace to be a fitting vessel to bear the saviour of the world; and, therefore, in view of the merits of her son Jesus Christ, Mary is conceived as integrally whole, ordered in herself and open to God to whom she turns and who makes it possible for her to do good, to love and speak the truth without imperfection. To put the *Immaculate Conception* negatively, Mary was conceived without original sin: without the tendency to sin transmitted to the whole human race through the process of human generation. In other words, Adam and Eve lost the integral gift of being created in a state of original justice – both good themselves and in a right relationship to God, to each other and to creation. Adam and Eve's original sin of disobeying the commandment of God was not only a personal sin but also, as our first parents, entailed the loss of the graced ordering of human nature which they had received from God. While we are given baptism, a trinitarian triple immersion into water and the mystery of the death and resurrection of Jesus Christ, an act which begins to heal the wound of original sin and with which we are called to cooperate, Mary was given a radical immersion in the redemptive grace of Jesus Christ from the first instant of her conception. Hence Mary was conceived both full of grace and without the wound of original sin.

With respect, then, to the theme of this final section on dogma and the Scripture, the dogma of the *Immaculate Conception* of Mary entails the view that she was conceived naturally, by her parents. In order for Mary to be conceived without original sin, which is transmitted as a deficiency of the original grace given to Adam and Eve and passed down through human procreation, then it follows that Mary must have been conceived, ensouled and given the grace of being immaculate from the first instant of her conception. If Mary was not given grace from the first instant of her conception then she would not have been perfectly free from original sin – because the deprivation of original sin is transmitted through the act of procreation. Therefore, the dogma of the *Immaculate Conception* is a certain interpretation of the moment of human conception, being the very first instant that a human being is one in body and soul; and, as such, gives a precise meaning to what the author of Genesis and Eve said: 'Now Adam knew Eve his wife, and she conceived and bore

Cain, saying, "I have gotten a man with the help of the Lord'" (Gn 4: 1).

The witness of Eve, as of so many others, is not just to a vague sense that God exists and acts; it is, rather, indicative of the whole biblical tradition that God is not defined abstractly – but acts in the concrete moments of existence. While, then, there are no doubt many questions about the beginning of creation and the beginning of each one of us, there is also the simple, literal truth, that where the human person comes to exist so God has acted to complete it.

Part IV

What is Certain and What is Uncertain about Conception

COMPRISING CHAPTERS FIVE, SIX AND SEVEN

What better starting point, then, than concrete and familiar phenomena and from there to go on to what is more difficult to understand but which belongs, as it were, to the same universe of truths? In other words, what is the meaning of conception, of human conception, if not the beginning of the life of the person; for, by definition of a person coming to exist, the person's coming to exist has a beginning.

Discussing the ordinary propagation of plants helps us to understand in a simple, concrete way, the natural order by and through which one kind of being reproduces another; and, in this simplistic sense, it anchors the imagination that can envisage a whole plethora of possibilities which, however, need a basis in reality if they are to be relevant. In other words, in view of the multitude of possibilities which an imagination can generate, itself going beyond the evidence speculatively and implying the existence of that which enables this to be possible, namely the soul informing the body and constituting the whole of human personhood – there is a need to think in terms of what actually exists.

As we have seen there are characteristic patterns of development where, for example, runner beans develop into runner bean plants and, in due course, produce runner beans for eating or planting. In other words there is a simple reality to there being, ordinarily, kind from kind; and, therefore, this is the case too when it comes to the transmission of human life: it is "life" from "life". Just as a bean is a bean and brings about the existence of plants and beans, so it is human beings that bring about human beings – but in this one respect differently, namely, through the action of God completing the whole by the integrated ensouling of the human soul.

There are, then, a number of certainties and uncertainties: Each one of us is a witness of having come to exist; and, having come to exist, having a beginning. However, there is no certainty, ordinarily, about which act of spousal love will contribute to the transmission of human life and draw from God, as it were, His ensouling action. At the same time as some may argue that we cannot be certain about whether or not we exist, whole and entire, from the very first instant of fertilization or not, we are certain that life is transmitted from the very first instant of fertilization. The sperm, as well as contributing its genetic inheritance and all that arises out of the specific characteristics of coming to rest in the ovum or egg, is at the same time totally energizing and sets in motion irreversible changes that are expressed in the closing of the now embryonic wall and the firing of the mitochondria and the driving of an uninterrupted process of development which, unimpeded, discloses by degrees the presence of the person[71]. The natural moment, then, for an absolute beginning to human life is the very first and irreversible moment of fertilization; and, therefore, this constitutes a nature sacrament: an outward sign of the inward action of God bringing the whole person to exist from the very first moment of fertilization. But, in view of the natural uncertainty as to whether or not a child has come to exist we are, in fact, called upon to be patient in our perception of the coming to be of a child. In other words, the

---

[71] Clearly, to suppose that there is no clear moment of beginning is more to do with the inadequacy of our perception than that there is not one; and, therefore, just as we take the outward sign of baptism, or any sacrament, as a sign of the inward action of God, so we can take what is clearly an outward sign of the beginning of human life as a definitive sign of the inward action of God – it not being possible to prove otherwise (cf. *Evangelium Vitae*).

natural uncertainty about a child coming to exist expresses, in its own way that each child is a gift; and, therefore, we are all a gift given to give thanks for the gift of life and to appreciate that we are all equally a gift.

Conversion to reality[72]

> It is integral to our human development to educate, to form (cf. CCC, 1783-1785) and to listen to the conscience (cf. CCC, 1779); and, as a part of that it is necessary to know and to recognize that 'Conscience *is not an infallible judge;* it can make mistakes'[73].

---

[72] Cf. "Is Faith Married Reason?", Chapter Ten: Part I (of *Volume III-Faith is Married Reason*, Newcastle upon Tyne: Cambridge Scholars Publishing, 2016. This paper, now included in the above book, was first presented as a response to an address by Bishop (now Cardinal) Angelo Scola; the Bishop's paper was called "The Nuptial Mystery at the Heart of the Church" (Oxford Catholic Chaplaincy, 21 March, 1998). The Bishop's paper is cited in this essay as: Angelo Scola, "The Nuptial Mystery" and page number.

[73] St. John Paul II, *Veritatis Splendor*, 62.

Watching plants grow stirs us to think about what is consistent in nature and, as such, leads us to consider the significance of this consistency; and, therefore, we can see that the reality of human beginning provides its own evidence to those who come into contact with the beginning of a human person.

In the following remarkable, and wonderful account of what took over seven years of pondering what was happening in the *in vitro* fertilization industry, we can see an embryologist discover that the reality of human embryonic development is a witness, in its own way, to the presence of the person from the beginning:

'I can remember always instinctively pausing each time my hand was over the biohazard disposal container, as if to double-check that I was, in fact, discarding the correct [human] embryos. Maybe it was also something else that caused me to pause.'

And he goes on to say, especially concerning the uncertainty around what was or was not a "viable" human embryo:

'Despite having some cells that are abnormal, an early embryo has the capacity to "self-correct." It does this by selectively pushing the abnormal cells out and replicating the normal cells. In fact, it appears that the abnormal cells become part of the placenta, leaving the normal cells to become the fetus. This is a remarkable biological process that is somehow "programmed" into these primitive cells.

To me [he says] this revelation reinforces the meaning of the Psalmist's words:

"For you formed my inward parts; you knitted me together in my mother's womb. I praise you, for I am fearfully and wonderfully made. Wonderful are your works; my soul knows it very well. My frame was not hidden from you, when I was being made in secret, intricately woven in the depths of the earth." Psalm 139:13-15'[74].

---

[74] *Dr. Craig Turczynski is a Reproductive Physiologist, Certified Teacher of the Billings Ovulation Method®, Director of Strategy and Scientific Affairs for BOMA-USA, and currently serves on the board of advisors for Sacred Heart Guardians and Shelter:* "Abnormal Embryos and Human Life: An

In the end, then, as with any inquiry, the inquirer has to address wholeheartedly and in all honesty the question of what, really, is the obstacle or obstacles to knowing the truth that each one of us has a beginning. May evidence and God Himself, the author of life, inspire whatever progress we may need to make to understand conception further, to recognize it more fully, and to rejoice more humbly in front of the amazing reality that we "behold". Let us remember, too, that the Lord wants no one to be lost; and, therefore, forgiveness awaits us all: He 'is forbearing

---

Embryologist's Post-Conversion Reflection": https://sacredheartguardians.org/abnormal-embryos-and-human-life/. Other testimonies can be found elsewhere: Laura Elm, "Foreword to Chapter Three" of *Mary and Bioethics: An Exploration*, by Francis Etheredge: https://enroutebooksandmedia.com/maryandbioethics/; and in *Within Reach of You: A Book of Prose and Prayers*, by Francis Etheredge, on p. 77, along with footnote: 27: "How many embryos are frozen in China's IVF clinics?" by Michael Cook, 24 Jan 2021: https://www.bioedge.org/bioethics/how-many-embryos-are-frozen-in-chinas-ivf-clinics/13677. *Within Reach of You* is published by En Route Books and Media: https://enroutebooksandmedia.com/withinreachofyou/.

toward you, not wishing that any should perish, but that all should reach repentance' (2 Peter, 3: 9).

Let us not forget, too, that the *word of God* complements science, just as faith and reason work together, being authored by the one God; however, in addition, the word of God has its own power to disrupt our settled errors[75] and to help us to see what is hidden from us, whether it be evidence or understanding.

---

[75] Cf. St. John Paul II, *Veritatis Splendor*, 63; cf. Cardinal Ratzinger, subsequently Pope Benedict XVI, "Conscience and Truth", *Communio*, 37, (Fall, 2010), p. 538; and cf. Pope Francis, *Gaudete et Exsultate,* 137, in his "Letter to Priests", https://www.hprweb.com/2019/09/letter-of-his-holiness-pope-francis-to-priests/#fnref-24676-26.

Chapter Five

The Teaching of the Church and the Problem of Uncertainty

Introduction: There are numerous directions that the exploration entailed in this book can go. However, in terms of fundamental points, the definition of conception is both required by a thorough understanding of the integrity of human being and, at the same time, extends throughout a whole range of issues: issues that, in the end, have a direct bearing on a person at the beginning of his or her life. The advantage, now, of discussing a specific legal case is that it summarises many of the problems surrounding a clear exposition of the beginning of the human person; but, at the same time, it shows that the implications of the truth unfold, ethically, throughout a whole range of modern developments and grounds, ultimately, in the possibility of a world-wide bio-legal declaration or law pertaining to the human race.

Prologue: A Modern Moment

Providence gives us many gifts and one of them is the help in a particular historical moment to develop our understanding of a truth that is related to our faith – even to a dogma of our faith: the moment of human conception; indeed, as the document on the Word of God, *Dei Verbum*, says:

> 'This tradition which comes from the Apostles develops in the Church with the help of the Holy Spirit[76]. For there is a growth in the understanding of the realities and the words which have been handed down. This happens through the contemplation and study made by believers, who treasure these things in their hearts (see Luke, 2:19, 51) through a penetrating understanding of the spiritual realities which they experience, and

---

[76] Footnote 5 in the document is: cf. First Vatican Council, Dogmatic Constitution on the Catholic Faith, Chap. 4, "On Faith and Reason:" Denzinger 1800 (3020); from: https://www.vatican.va/archive/hist_councils/ii_vatican_council/documents/vat-ii_const_19651118_dei-verbum_en.html.

through the preaching of those who have received through Episcopal succession the sure gift of truth. For as the centuries succeed one another, the Church constantly moves forward toward the fullness of divine truth until the words of God reach their complete fulfilment in her' (8).

In other words, the reality of participation in the development of our faith is multi-personal and multi-disciplinary; and, in addition, there is progress in the truth which brings faith and reason into a renewed dialogue: a dialogue that has, in some way, to go beyond recriminations, regrets and even accusations[77]. For, bearing in mind the enormous implications of how the world has lived in the light of the uncertainty about the beginning of human life, we need to proceed with an amazing interrelationship of truth and love. But, at the same time, we are on the threshold of untold developments in which what was thought to be an ethical development is in fact in danger of being the

---

[77] Cf. Pope Francis, *Fratelli Tutti*, articles 226-227 : https://www.vatican.va/archive/hist_councils/ii_vatican_council/documents/vat-ii_const_19651118_dei-verbum_en.html.

total opposite. Thus there are frozen human embryos which have not only suffered the real frustration of the right to natural, completing human development, but which are now being proposed for some kind of mass production whereby they have been 'prevented from reaching their full potential by depriving them of the extra-embryonic cells required for implantation into the uterus, [whereas] … if they, along with the extraembryonic cells, were implanted into a surrogate uterus, [they] could develop into a living human infant'[78]. Therefore, there is an urgency to clarify what we now understand to be the moment of human conception: a conception which establishes the human relationships regulated by human rights; for, not only are we conceived-in-relationship but we are conceived in a relationship of reciprocal rights from conception onwards.

---

[78] "Mass Production of Human "Embryoid" Cells from Developmentally Frozen Embryos: Is it Ethical?": http://www.cmq.org.uk/CMQ/2020/Aug/mass_production_of_embryoid_cell.html.

Introduction: Who is My Neighbor?

The *Congregation for the Doctrine of the Faith* said: 'Beginning at conception, children suffering from malformation or other pathologies are *little patients* whom medicine today can always assist and accompany in a manner respectful of life. Their life is sacred, unique, unrepeatable, and inviolable, exactly like that of every adult person' (the *Good Samaritan, Samaritanus Bonus,* 6[79]).

Why, however, begin a discussion on the nature of conception with a reference to a document which, generally, attends to the sensitive care and treatment, as well as the accompaniment of, people approaching the end of life and the help that they need: particularly the help that they need to hope in God and to experience the love of neighbour, whether that neighbour is a member of his or her family, a doctor, a nurse or any person with whom they come into contact? Indeed, 'neighbour' is a word rich in

---

[79] Congregation for the Doctrine of the Faith: https://press.vatican.va/content/salastampa/en/bollettino/pubblico/2020/09/22/200922a.html.

significance[80]: 'neighbor (n.): "one who lives near another," Middle English neighebor, from Old English ... "one who dwells nearby," from neah "near" ... "dweller," related to bur "dwelling," from Proto-Germanic * ... "to be, exist, grow".' In other words, there are many who live near another, dwell nearby and are, whether literally near or close to the heart of others in need, a neighbor: a concrete help to others. Who, then, can be a closer neighbor to the child conceived than God Himself; and, therefore, who better to communicate the truth concerning conception than the Church? As *Gaudium et Spes* says: 'For by His incarnation the Son of God has united Himself in some fashion with every man' (22[81]); and, therefore, if this is a union prior to baptism what moment is prior to that of conception and corresponding, as it were, to the moment of the *Incarnation*?

---

[80] The definition of neighbour was provided by Mr. Martin Higgins, MA, an Eastern European Linguist.

[81] http://www.vatican.va/archive/hist_councils/ii_vatican_council/documents/vat-ii_const_19651207_gaudium-et-spes_en.html.

The Problem of Uncertainty in Both Church Teaching and the "Opinion of the Court" in Roe v Wade

But to go back to the question: Why, however, begin a discussion on conception with reference to a document on hope and compassion for the vulnerable, whether elderly or just conceived? There are two, if not three reasons for beginning in this way. Firstly, the compassion and hope which fills this document is as applicable, as it suggests, to the needs of the child conceived, their mothers and fathers, as to the needs of the vulnerable at whatever time and stage of life; indeed, especially where 'Beginning at conception, children suffering from malformation or other pathologies are *little patients* whom medicine today can always assist and accompany in a manner respectful of life' (*Samaritanus Bonus*, 6). Secondly, yet again the Church has used the expression, concerning the beginning of human life: 'Beginning at conception, children ...' (*Samaritanus Bonus*, 6); and, therefore, the Church is clearly articulating a constant teaching that a child is conceived at conception. In the *Gospel of Life*, Pope St. John Paul II says: Every person can come to recognize 'the sacred value of human life from its very beginning until its end, and can affirm the right of every human being to have this primary good respected to the highest degree' (2[82]); and, indeed, what does conception ordinarily mean but the 'very beginning'? *Donum Vitae*, on the *Gift of life*, says in its introductory comments: that *'the first part will have as its subject respect for the human being from the first moment of his or her existence'*[83]; and, therefore, can he or she exist from the first moment of his or her existence without being a person? It is certainly true that a person can exist without being recognized to be present; and, as such, there is a process through which that presence, which is ordinarily hidden, is made increasingly visible and recognizable. Thus, as even the Warnock Report admitted, there is ordinarily a seamless unfolding of each one of us from conception onwards. Thirdly, again the Church speaks of the child's life being 'sacred, unique, unrepeatable, and inviolable, exactly like that of every adult

---

[82] *Evangelium Vitae*: http://www.vatican.va/content/john-paul-ii/en/encyclicals/documents/hf_jp-ii_enc_25031995_evangelium-vitae.html.

[83] *Donum Vitae*: https://www.vatican.va/roman_curia/congregations/cfaith/documents/rc_con_cfaith_doc_19870222_respect-for-human-life_en.html.

person' (*Samaritanus Bonus*, 6); and, as such, echoes Pope Paul VI and Pope John XXIII, who said: "'Human life is sacred—all men must recognize that fact," Our predecessor Pope John XXIII recalled. "From its very inception it reveals the creating hand of God"[84]'.

The Problem of Uncertainty in both Church Teaching and the 14th Amendment

On the one hand in 1995, in *Evangelium Vitae, The Gospel of Life*, Pope St. John Paul II quotes from the Declaration on Procured Abortion, published in November, 1974, which said: 'modern genetic science ... has demonstrated that from the first instant there is established the programme of what this living being will be: a person, this individual person with his characteristic aspects already well determined' (60)[85]. The phrase which attracts

---

[84] *Humanae Vitae*, 13: http://www.vatican.va/content/paul-vi/en/encyclicals/documents/hf_p-vi_enc_25071968_humanae-vitae.html.

[85] *Evangelium Vitae*: http://www.vatican.va/content/john-paul-ii/en/encyclicals/documents/hf_jp-ii_enc_25031995_evangelium-vitae.html.

attention, in this context, is where the declaration speaks of what has been established: 'modern genetic science … has … established … what this living being will be: a person'. In other words, 'modern genetic science' has established the possibility that there 'will be' a person; and, if there 'will be' a person, then when will that person be there? When will the person who will come to exist – come to exist from the existence of 'the living being' which does exist? Thus there seems to be a distinction between the 'living being' which has come to exist and the person which will come to exist. What has come to exist, then, that is a 'living being' and yet is not a person? As a kind of explanation of the possible thinking behind this distinction between a 'living being' and the person that that 'living being will be' there is a note that was added to the English translation of *Donum Vitae, The Gift of Life*.

> 'The point of this part of the discussion is to show that, in fact, there was a dimension of meaning that "seemed to be omitted by addition" in the *English* translation of *Donum Vitae*, namely, that the zygote comes to exist 'when the nuclei of the two gametes have fused'. Thus the additional English expression, 'the nuclei of', is both an additional phrase to what is in the Latin text

and, at the same time, it is a phrase that seems to delimit the definition of a zygote to the fusion of the two nuclei. However, as we shall see, the Latin expression, *'orta a fusione'* (arising from a fusion), does not mention nuclei and, therefore, is a more comprehensive account of the nature of the zygote. Thus the Latin text may include, in its range of meaning, the development of the zygote from the first instant of fertilization; and, therefore, the Latin may well be more inclusive of that first instant than the English expression: 'when the nuclei of the two gametes have fused'. The expression, 'when the nuclei of the two gametes have formed' tends to make one think that the zygote has not formed until the nuclei have fused; and, as such, could be described as an interpretative translation, referring to a developmental point which is not so clearly evident from the Latin text. For there is a case to advance that the Latin expression, *'orta a fusione'* (arising from a fusion), could apply from the very first moment that fusion occurs: the first instant of fertilisation'[86].

---

[86] Excerpt from Francis Etheredge, *The Human Person: A Bioethical Word*, p. 369, St. Louis: En Route Books and Media, 2017.

On the other hand, on January 22, 1973, the U.S Supreme Court legalized abortion in Roe v. Wade; and, without going into the details of the whole judgement, pronounced that the 'State ... has legitimate interests in protecting both the pregnant woman's health and the potentiality of human life'[87]; however, the wording of the "Opinion of the Court" is inconsistent: it both refers to the 'potential life' and 'fetal life after viability'[88]. In other

---

[87] Roe v Wade: https://caselaw.findlaw.com/us-supreme-court/410/113.html.

[88] 'With respect to the State's important and legitimate interest in potential life, the "compelling" point is at viability. This is so because the fetus then presumably has the capability of meaningful life outside the mother's womb. State regulation protective of fetal life after viability thus has both logical and biological justifications': p. 163 of ROE v. WADE Syllabus ROE ET AL. v. WADE, DISTRICT ATTORNEY OF DALLAS COUNTY APPEAL FROM THE UNITED STATES DISTRICT COURT FOR THE NORTHERN DISTRICT OF TEXAS No. 70-18. Argued December 13, 1971-Reargued October 11, 1972-Decided January 22, 1973. Pp. 113-178: https://tile.loc.gov/storage-services/service/ll/usrep/usrep410/usrep410113/usrep410113.pdf.

words, it looks as if the Court's opinion is that there is 'potential life' up until there is 'fetal life after viability'. But, one may ask, what is the woman pregnant with? If she is pregnant with the 'potentiality of human life' then what is that potentiality except for viability? - for viability refers to the child being able to live outside the womb, with help, just as the womb is the right environment for the child to develop in. Thus the "Opinion of the Court" has conflated the existence of a child's life and viability whereas viability is only possible because of the existence of a child's life in the first place; and, therefore, the very recognition of viability is recognition of the life of a child from conception. The view that a child's life is not a child's life because it is not developed enough to exist outside of the womb is the same as claiming that any child that cannot live independently is not a child; but, in reality, it is in the very nature of being a child that development proceeds from conception onwards, through all the development stages that characterize a person's life, and which include pre-and-post birth stages of development. Thus the claim that a child is not a child if it is not possible for it to survive outside the womb is an incoherent and contradictory claim which has nothing to do with the reality of human development and everything to do with a prior judgement of a philosophical nature: a claim that is neither vindicated by biology nor logic. The claim that a child is not, by definition of his or her very existence, a human child, is contradicted by the very identity of the sperm and the egg through which he or she came to exist or, in the case of technological interventions, the human egg and nucleus through which he or she came to exist[89].

Furthermore, the claim that a child is not, by definition of his or her very existence, a human child, is not a logical claim; for, if what begins is typically human then he or she has a characteristically, developmentally, progressive expression of his or her identity. Therefore it follows that what has begun is typically the first of many stages through which his or her development passes in the course of showing "who" was present from the beginning; and, if this is true, then it follows that we are obligated to recognize the rights of this child: to life; to completing human development; and to integrity – to begin with but a few that need to be recognized. For, more generally, if the truth establishes our common human identity then it

---

[89] Cf. Profs. Justo Aznar and Julio Tudela: Chapter 5: Part II of *Conception: An Icon of the Beginning*, for a biological account of the human identity of the human embryo.

establishes, at the same time, our ethical responsibility for each other.

Potential life could be a quasi-philosophical concept which makes an unfounded distinction between being a potential human being and viability: as if viability introduces something that was not there before viability.

Alternatively, it looks as if the Court's opinion is that there is 'potential life' up until there is 'fetal life after viability'. Is, then, the "Opinion of the Court" saying that prior to 'viability' there is the potential for 'viability'? However, the potential for viability entails that a child is developmentally on course to becoming viable. It may be the case that 'viability' is, in the context of the "Opinion of the Court", relevant because the terms of the 14th Amendment refers to legal citizenship commencing, as it were, with birth (or naturalization); and, therefore, there may be the implicit claim that the 'viability' of the child brings him or her within the perceived jurisdiction of the 14th Amendment. For if the viability of the child is equivalent to surviving premature birth, being born prematurely nevertheless brings a person into American citizenship. Hence the implication that abortion is not the deliberate killing of an American citizen if the child is not 'viable' –

but only expresses a potentiality for viability outside the womb.

Viability, however, is a stage in the normal development of a child; and, therefore, the abortion of a child prior to his or her viability is clearly discriminatory and needs redress. According to the rule of law that redress could be according to how the right to life falls under the jurisdiction of a Constitution[90] or it could be more widely addressed on the basis of bio-legal principles that can inform, not just national governments but the good of the human race[91]; indeed, in a certain sense, progress in the natural justice of a Constitution, or a law is progress towards the possibility of good international bioethical law.

In the end, then, aside from the possibility that viability is here being understood as determining whether or not

---

[90] From Ed Whelan, "Judge Barrett on Stare Decisis": https://www.nationalreview.com/bench-memos/judge-barrett-on-stare-decisis/.

[91] Cf. "Principles of international biolaw: Seeking common ground at the intersection of bioethics and human rights": Roberto Adorno, Bruylant, 2013: "Chapter 1: Principles of International Biomedical law", pp. 13-35: https://www.academia.edu/4063596/Principles_of_international_biolaw.

the Constitution can be said to apply to a viable child, there is also the possibility that there seems to be a coincidence of meaning between one reading of the 'potentiality of human life' and the expression used in a document of the Catholic Church: that a woman is pregnant with a 'living being' that 'will be' a person or the woman is pregnant with whatever it is that has the 'potentiality of human life' but, again, is not yet a person.

In the first case there 'will be' a person and in the second case there is the 'potentiality of human life' and, presumably, at some indefinable point a human person. In the context of each document, however, there is a world of difference between how these two starting points are evaluated. In the case of *The Gospel of Life*, Pope St. John Paul II says: 'what is at stake is so important that, from the standpoint of moral obligation, the mere probability that a human person is involved would suffice to justify an absolutely clear prohibition of any intervention aimed at killing a human embryo' (*Evangelium Vitae*, 60). Whereas in the case of Roe v Wade the uncertainty as to what the woman is pregnant with has led to the possibility of the destruction of whatever she is pregnant with which has the 'potentiality of human life'. In other words, there is uncertainty in both expressions of what has come to exist at conception and yet they are evaluated in completely different ways.

What follows next is a discussion of some aspects of the "Opinion of the Court" of Roe v Wade and the timely necessity of the development of Church teaching. Thus it is now necessary to examine, more closely, the issues that arise from an examination of the "Opinion of the Court" and a dissenting opinion in Roe v Wade.

CHAPTER SIX

ON THE INTERPRETATION OF TEXTS: PARTICULARLY THE 14TH AMENDMENT

AMENDMENT 14

'[1.] All persons born or naturalized in the United States, and subject to the jurisdiction thereof, are citizens of the United States and of the State wherein they reside. No State shall make or enforce any law which shall abridge the privileges or immunities of citizens of the United States; nor shall any State deprive any person of life, liberty, or property, without due process of law; nor deny to any person within its jurisdiction the equal protection of the laws'[92].

As, however, the 14th Amendment states that it is concerned with 'All persons born or naturalized in the United States, and subject to the jurisdiction thereof, are citizens of the United States and of the State wherein they reside'

---

[92] https://www.aclu.org/united-states-constitution-11th-and-following-amendments#14.

it could, therefore, be said to exclude a more universalist claim to apply to all human beings, whether in America by immigration or for whatever other reason. But then it also says: 'nor shall any State deprive any person of life, liberty, or property, without due process of law; nor deny to any person within its jurisdiction the equal protection of the laws'[93]. In other words, there is scope for the possibility that the meaning of the statement that no State can 'deprive any person of life' is in fact of a more universal character; and, therefore, there is a sense in which, perhaps not fully consciously, the legislators intended an implicit declaration of human rights which, in the nature of the law, is simply expressed according to the natural jurisdiction of this law applying specifically to those born or naturalized in America[94]. In other words, as citizenship

---

[93] https://www.aclu.org/united-states-constitution-11th-and-following-amendments#14.

[94] My perception of this point was sharpened by my daughter, Grace Etheredge's, reading of a draft of this article; and, in addition, the recognition that there will be a legal tradition defining the term 'person'. But, at the same time, it must be understood, too, that meaning is never confined to one expression of it and, therefore, there are wider considerations that

Chapter Six: On the Interpretation of Texts   183

naturally applies to those 'born or naturalized in America', it follows that the American Constitution is not excluding the right to life from conception onwards; rather, the Constitution is defining the common understanding of an American citizen being, ordinarily, one 'born or naturalized in America'. Therefore, as we shall see in the following paragraphs, the right to life is within the general understanding of being 'born or naturalized in America' as, without life, there is no child to be 'born' an American citizen or, subsequently, no person to be naturalized.

What, then, will help with the interpretation of legal texts?[95] Thus, just as there is a concern for the original

---

are needed to help us to understand the use of any specific, legal terminology.

[95] See: "9 Things You Should Know About Supreme Court Nominee Amy Coney Barrett", September 28, 2020, by Joe Carter: https://www.thegospelcoalition.org/article/9-things-amy-coney-barrett/:

'5. In her judicial philosophy, Judge Barrett is considered a proponent of originalism, a manner of interpreting the Constitution that begins with the text and attempts to give that text the meaning it had when it was adopted, and textualism, a method of statutory interpretation that relies on the plain text of a statute to determine its meaning.

meaning of the philosophical understanding of 'first matter' and 'form'[96] which lies behind the triple conception of human being, whether in Aristotle or St. Thomas following him, so there is an interpretation of legal texts which endeavours to understand the original intention of the document. Just as there are principles and practices basic to the interpretation of Scripture which, generally, develop from the principle that the literal sense is the sense intended by the author[97], so there seems to be a parallel sense of interpreting legal texts: 'a method of statutory interpretation that relies on the plain text of a statute to determine its meaning'[98].

---

[96] 'Form' is here understood as that which determines the pure potentiality of 'first matter' to be a specific entity e.g. plant, animal or rational being.

[97] Cf. Etheredge, *Scripture: A Unique Word*, Cambridge Scholars Publishing, 2014.

[98] Joe Carter: 5th of "9 Things You Should Know About Supreme Court Nominee Amy Coney Barrett".

Clearly, then, the American Constitution follows Lincoln's address that 'all men are created equal'[99]; for, in the common understanding of being created equal, there is an ordinary sense, no doubt common at the time, that whatever constituted the original moment of a human being's creation was a moment common to all. Therefore, it could be said, whatever is now understood to be the common starting point of all human beings is what establishes our common equality before the law. Thus the principle: Where the body lives, there the soul is and where both are is the person[100]. Therefore, the first instant of human conception is ordinarily the first moment of a person beginning to exist. Where this first instant is subsequent to the first instant of fertilization, as in the case of twins, the first instant that the body of the twin comes to exist is the first instant that the twin human being has come to exist; and, irrespective of the difficulties of conjoined twins, the reality of conjoined twins shows forth the truth that bodily existence expresses personal existence. In the first instant

---

[99] Dr. Elizabeth Rex, "End Word", p. 607 of *Conception: An Icon of the Beginning*.

[100] See this theme developed throughout *Conception: An Icon of the Beginning*.

of what is established by artificial methods of human conception, it is the first instant in which the bodily existence comes to exist that there is what expresses the existence of the human person, whole and entire; it being the nature of bodily development to progressively manifest the existence of the human person from conception[101].

In other words, again in the common understanding of the time, a person would have been understood to exist from conception in that the life of the person exists from conception: parents conceive children not plants and mourn the loss of a child – not the loss of a plant; and, now

---

[101] Tina Beattie, "Catholicism, Choice and Consciousness: A Feminist Theological Perspective on Abortion", on p. 53: She says: 'I avoid scientific debates about embryonic development …' but then goes on to claim, on p. 60: 'There is no 'very first instant' of human existence, because the process between fertilization and implantation of the fertilized zygote takes several days.' However, without a 'first instant' there is no subsequent development; and, moreover, all subsequent development is continuous from the first instant. Therefore Beattie's claim is not only unscientific but incoherent. Personhood cannot be absolutely ruled out and, for the good reasons of this booklet, is thrice argued to be from the first instant of human conception: scientifically; philosophically; and theologically.

Chapter Six: On the Interpretation of Texts

that conception is better understood to have a first instant, then it follows that personhood begins from the first instant of human conception: from the first instant that the sperm-egg union effects the whole of the embryonic child expressed, bodily, in the enclosing of the sperm in the newly formed and active embryonic wall. In other words, there is a moment when there is an active sperm and an inert egg and then there is the embryonic child expressed in the bodily integrity of the newly "walled" human embryo. For, quite apart from the philosophical problems that have arisen and, in a sense, which have always accompanied the definition of terms, there is the reality of a living presence from the first instant of conception: a person's life: what is understood to be the life of a human child: the life of a boy or a girl. The American Constitution, then, establishes a wonderful precedent to which not only American abuses of this truth can be appealed[102],

---

[102] Whether it be the abolition of slavery (Rex, "End Word" p. 607 of *Conception: An Icon of the Beginning*), the mistreatment of people suffering from syphilis (cf. Etheredge, *The Human Person: A Bioethical Word*, "General Foreword" by Dr. Mary Anne Urlakis, pp. 13-16) or the manipulation of ignorance which led to such tragic experiments on women in the

but to which the world can "measure" its juridical claim to expressing and embodying equality before the law; indeed, while other legislation is more explicit, such as the 1990 German Embryo Protection Act passed 'in compliance with the Nuremburg Code'[103], the American Constitutional reference to the life of a person is as true now as it always was and communicates a common understanding that each one of us begins at conception.

Nevertheless, the opening wording of the 14th Amendment provides a context in which to understand the protection of the life of a person; for, it says: 'All persons born or naturalized in the United States, and subject to the jurisdiction thereof, are citizens of the United States'. In other words, a person is clearly understood to refer to 'All persons born or naturalized'; and, in general, a number of the cases which were heard with respect to this Amendment were concerned with, quite rightly, the reality of

---

trials of contraceptive pills (Etheredge, footnotes 780-781 on pp. 583-584 of *Conception: An Icon of the Beginning.*

[103] Pp. 597-598 of *Conception: An Icon of the Beginning*, Elizabeth Rex's "End Word".

Chapter Six: On the Interpretation of Texts   189

equality between races before the law[104]. Thus, it could be argued, the 14th Amendment is referring to the life of a person who is born, in that this simply reflects the concern of the legislation at the time, bearing in mind its desire to establish racial equality before the law; however, the life of a person who is born may well be a way of presuming that the life of a child, prior to birth, is implicitly defended in that this naturally leads to the life of 'All persons born'. In other words, the fact that the words of the 14th Amendment are concerned with the life of a person who is born is naturally the basis on which to recognize the implicit concern of protecting the life of all innocent people – from conception onwards.

One of the ways, then, of corroborating this claim is looking at the cultural climate which developed in the course of the period both leading up to and following the 14th Amendment. In other words, there are cultural, even

---

[104] Cf. the "List of 14th Amendment Cases": https://en.wikipedia.org/wiki/List_of_14th_amendment_cases; but, as one can see, there are many more cases that need to be considered as falling under this Amendment: https://en.wikipedia.org/wiki/Fourteenth_Amendment_to_the_United_States_Constitution.

legal developments, which elucidate the basic meaning of a text. In the case of the contemporary question of the validity of Anglican Orders, the question of what intention was expressed in the Anglican ordination of a priest, it was historically demonstrated that an Anglican priest was ordained in a specifically different sense to that of his contemporary, Catholic priest. In other words, an Anglican ordination was specifically intended to repudiate, or reject, the Catholic understanding of the priest as celebrating, effectively, a sacramental mystery of the presence of the body and blood of Jesus Christ[105]. The point of this observation, however, is not to confuse the issue by assuming that there is no other intention possible, either then or now, but rather to recognize that a historical account can elucidate what is specific in a public enactment

---

[105] Cf. For example, a concise summary of the point in question raised and explained by Dr. Francis Clark, SJ, in *Anglican Orders and Defect of Intention*. By Francis Clark, S.J. Pp. xx + 215. London: Longmans, Green, 1956. 25s. A. R. Vidler [a1] https://www.cambridge.org/core/journals/journal-of-ecclesiastical-history/article/anglican-orders-and-defect-of-intention-by-clarkfrancis-sj-pp-xx-215-london-longmans-green-1956-25s/F80F57967A7D266CC9F7F2EDBE849D92.

characteristic of a historical period. Similarly, then, the very developing of laws, across the United States[106], protecting a child from the possibility of *abortion is itself confirming evidence of the historical intentionality expressed in the 14*th *Amendment.* Indeed, this is precisely the argument of the dissenting judge in Roe v Wade:

'MR. Justice Rehnquist, dissenting

'As early as 1821, the first state law dealing directly with abortion was enacted by the Connecticut Legisla-

---

[106] Cf. Christian Myers, "Law Professor Reflects on Landmark Case": 'Barrett also outlined the history of the Roe v. Wade decision and associated cases in the Supreme Court.

At the time of the case, most states prohibited abortion, except in cases wherein it protected the life of the mother, she said': https://ndsmcobserver.com/2013/01/law-professor-reflects-on-landmark-case/; and, more generally, the article by Steven Mosher: "How Amy Coney Barrett will use science and legal principles to overturn Roe v. Wade": https://www.lifesitenews.com/blogs/how-amy-coney-barrett-will-use-science-and-legal-principles-to-overturn-roe-v-wade.

ture'[107] ... 'By the time of the adoption of the Fourteenth Amendment in 1868, there were at least 36 laws enacted by state or territorial legislatures limiting abortion.' In due course Rehnquist says: 'The only conclusion possible from this history is that the drafters did not intend to have the Fourteenth Amendment withdraw from the States the power to legislate with respect to this matter'[108].

In other words, both before and after the 14th Amendment the very development of legislation which protected

---

[107] Pp. 174-175 of pp. 113-178: ROE v. WADE Syllabus ROE ET AL. v. WADE, DISTRICT ATTORNEY OF DALLAS COUNTY APPEAL FROM THE UNITED STATES DISTRICT COURT FOR THE NORTHERN DISTRICT OF TEXAS No. 70-18. Argued December 13, 1971-Reargued October 11, 1972-Decided January 22, 1973: https://tile.loc.gov/storage-services/service/ll/usrep/usrep410/usrep410113/usrep410113.pdf.: the precise law was included in the text but here is in this footnote: 'Conn. Stat., Tit. 22, §§ 14, 16.'

[108] P. 177 of pp. 113-178 of ROE v. WADE Syllabus ROE ET AL. v. WADE etc.

Chapter Six: On the Interpretation of Texts

the life of the child, as he or she became increasingly threatened due to the increasing practice of abortion *is itself evidence of how the 14th Amendment's protection of the life of the person was and is to be understood.* In other words, while the 14th Amendment was specifically addressing the rights of citizens, the more universal right to life is implicated; and, as such, the 14th Amendment understood its remit to be that of referring to legal citizenship and, as such, no doubt had in mind the common practice of registering a birth, or the naturalization of a citizen, – as the natural point from which citizenship dates. Citizenship, however, registers a workable legal definition of a subject of the American Constitution; and, therefore, citizenship is the natural expression of what follows on a human life *as it becomes possible to identify the application of citizenship.* Thus, for example, a premature baby falls quite squarely within the meaning of the birth of an American citizen; and, even recently, it is furthermore recognized in President Trump's specific intention to sign a "The Born-Alive Infant Abortion Survivors Act" which will ensure medical treatment of a child that survives an abortion[109]. The question, then, of citizenship is different from the question of preserving the life of a child; but, clearly, they are related: if the life of a child is not protected from conception then there is not a citizen to be legally recognized on being born and registered. The recognition of potential citizenship is different, then, to the speculative claim of a 'potential life': the former is a coherent legal definition whereas the latter is both unverifiable and derived, it seems, from a quasi-philosophical understanding of the relationship of a child's viability to a right to life.

On the Question of the Rightful Protection of Women

As regards the related claim that pro-abortion legislation is about protecting women[110] it is true that, in

---

[109] Christine Rousselle, "Trump announces 'Born Alive' executive order for abortion survivors": https://www.catholicnewsagency.com/news/trump-announces-born-alive-executive-order-30262.

[110] There are various possible references for this discussion, around pages 140-160 of 113-178 of ROE v. WADE Syllabus ROE ET AL. v. WADE etc.

general, regulating a procedure entails protecting a person from harm; and, indeed, in view of an abortion that is almost equivalent to self-harm, there is indeed the possibility of a great injury that the woman, or someone else, inflicts on her – but it is not on her alone. Nevertheless, this tragic response to being pregnant is not answered by "medicalizing" abortion itself – as if it solely about protecting the woman. In other words, it is far from established that there is no harm to the woman, *per se*, in virtue of the very act of carrying out even a medicalized, or elective abortion. Indeed, the many testimonies of women who have had abortions and profoundly regretted them[111], quite apart from increasing psychiatric evidence of an adverse psychological reaction to what was thought to be some kind of simple "medical procedure"[112], not to

---

[111] Why is this kind of evidence not taken into account? See: http://abortionmemorial.com/; indeed, it is increasingly recognized that the father, as well as other members of the family, are suffering too (cf. also: https://foundationsoflife.org/shes-pregnant-what-do-i-do-now/.

[112] Cf. "Post Abortion Syndrome": https://thelifeinstitute.net/learning-centre/abortion-effects/post-abortion-syndrome.

mention the various medical complications of such an act[113], all contribute to the view that an elective abortion is not in the interest of the health of the woman. Therefore, whatever distressing circumstances there may be involved in the actual context of a woman deciding on an abortion, it is clear that an abortion is an invasive medical action on "who" is not part of the woman's body; for, if the child were a "part of the woman's body"[114] then the same relationship would exist between her and her own body, whether it be a natural part of her like a limb or a growth, like a cancer. But, as we know, the loss of a limb is a different kind of grief to the loss of a child. If, in other

---

[113] Cf. "Risks about Abortion": https://foundationsoflife.org/facts-about-abortion/risks-about-abortion/.

[114] 'In fact, it is not clear to us that the claim asserted by some amici that one has an unlimited right to do with one's body as one pleases bears a close relationship to the right of privacy previously articulated in the Court's decisions. The Court has refused to recognize an unlimited right of this kind in the past': p. 154 of 113-178 of ROE v. WADE Syllabus ROE ET AL. v. WADE etc. Where is the proof that carrying a child falls under the terms of the following?: a 'right to do with one's body as one pleases'.

words, there is no child present, then the medical action would be an amputation or the removal of a random growth, and not an abortion. In a word, then, the direct object of an elective abortion is the removal of a child that a woman is carrying; and, at the same time, the removal of this child is generally done in such a way as that it does not involve the least consideration for preserving the life of that child. Deliberate abortion, then, for whatever reason, entails the implication of a "relationship" to the child, not just lost but removed, and this unaddressed relationship, whether denied or showing itself in unexpected ways, needs ultimately to be addressed by both the child's mother and father.

By implication, then, a medical act that intends to help a child to be better placed for his or her development and, at the same time, is an act by which the mother is helped, is clearly of a different kind; and, therefore, one wonders why, given the increase in medical expertise, why there are so few documented cases of a woman being helped with an ectopic pregnancy: a human embryo that implants in the fallopian tube or in some other place rather than in the womb. In other words, does the overriding mentality of abortion and abortifacient drugs obscure the real help that both mother and child need?

In sum, having read through the "Opinion of the Court" in Roe v Wade there is neither any real recognition of the harm to the woman of the event of the abortion itself nor any mention of the obvious harm to the child: the truth about what actually happens to a child in an elective abortion. However, reviewing the "Opinion of the Court" has, very helpfully, brought out a number of aspects that, as we can see, need a resolution.

The Principle of Determining an Appropriate Level of Legal Action

> 'Barrett … states that she "tend[s] to agree with those who say that a justice's duty is to the Constitution and that it is thus more legitimate for her to enforce her best understanding of the Constitution rather than a precedent she thinks clearly in conflict with it. That itself serves an important rule-of-law value"'[115].

---

[115] From Ed Whelan, "Judge Barrett on Stare Decisis": https://www.nationalreview.com/bench-memos/judge-barrett-on-stare-decisis/.

Chapter Six: On the Interpretation of Texts

In a word, the historical consistency with which a statute is interpreted, whether in a constitution or elsewhere, reflects its underlying meaning and is a part of the evidence of what constitutes its original and stable meaning; and, therefore, the evidence of legislation, pre and post 14th Amendment, protecting human life against a rising tide of claims to justify deliberate abortion is itself evidence of what was understood by the 14th Amendment's protection of the life of the person. Furthermore, then, in the specific context of the American Constitution it makes sense that the first duty of a judge is to express 'her best understanding of the Constitution'; and, at the same time, it makes sense to have a view as to which level of the judiciary or the legislature is appropriate for the handling of a particular case or enunciating a particular law for the good of both *the rule of law* and *the good government of a country*[116]. More widely, then, this raises the possibility of bio-

---

[116] MICAIAH BILGER: Amy Barrett Believes Life Begins at Conception, Questions Roe's "Judicial Fiat" of "Abortion on Demand":

https://www.lifenews.com/2020/09/25/amy-barrett-believes-life-begins-at-conception-questions-roes-judicial-fiat-of-abortion-on-demand/.

legal principles that can inform, not just national governments but the good of the human race[117].

---

[117] Cf. "Principles of international biolaw: Seeking common ground at the intersection of bioethics and human rights": Roberto Adorno, Bruylant, 2013: "Chapter 1: Principles of International Biomedical law", pp. 13-35: https://www.academia.edu/4063596/Principles_of_international_biolaw.

CHAPTER SEVEN

AN ANSWER TO THE UNCERTAINTY OF WHAT OR WHO EXISTS FROM CONCEPTION

Uncertainty is a characteristic of human experience; and, therefore, it is not unusual for a variety of factors to help determine the reality of what actually exists: What the real situation actually is. Uncertainty, however, expresses a value. There is an uncertainty about whether or not a specific act of spousal love will beget a child; and, indeed, it is possible that the very uncertainty that exists allows for the perception and reception of a child as a gift: a gift from God[118]. In the case of pregnancy, then, these two very different documents agree that there is an uncertainty about what is happening. On the one hand the woman is pregnant with a 'living being' and, on the other hand, with what has a 'potentiality of human life'. In neither case does there seem to be any clarity about what comes to exist at conception; except, that is, there is a common agreement that "something" comes to exist at

---

[118] Cf. Etheredge, *The Human Person: A Bioethical Word*, St. Louis: En Route Books and Media, 2017: pp. 61-62.

conception. The question is, then, what comes to exist at conception. What follows are a number of considerations which, together, constitute an answer to the need for certainty.

What is the Experience of Women in Pregnancy?

My wife spoke of looking forward to meeting the person conceived[119]; and, as such, echoed the certainty of Eve: 'I have gotten a man with the help of the Lord' (Gn 4: 1; but consider the whole biblical witness to the real experience of women, children and men written about in the Scriptures). Even, then, if it was not so clear to my wife that conception involved an act of God, it was certainly clear to my wife that she had conceived a child: 'a man' (Gn 4: 1). There are many other testimonies, too, both in

---

[119] This was put more formally in the article, Francis Etheredge, "The Mysterious Instant of Conception": *The National Catholic Bioethical Quarterly*, Autumn 2012: https://www.pdcnet.org/C1257D43006C9AB1/file/11DAAEF87F30A61985257D6D00682134/$FILE/ncbq_2012_0012_0003_0041_0050.pdf.

Chapter Seven: An Answer to the Uncertainty

human experience generally[120] and in the human experience embedded in the Scriptures. What is the identity of the human longing for a child which makes the suffering of infertility so painful?[121] What is so disappointing about a miscarriage? Surely it has nothing to do with the abstract claim of losing a blob of cells: a claim so contrary to the reality of the organized human embryonic child whose development is so interactively ordered to his or her presence in the nurturing womb of his or her mother[122]. What is the value of this human experience?

---

[120] Cf. Etheredge, *The Prayerful Kiss,* St. Louis: En Route Books and Media, 2019: the poem and prose entitled "Indelible": an account of the loss of a child to abortion from the experience of a father.

[121] Cf. Etheredge, *Mary and Bioethics: An Exploration*, St. Louis: En Route Books and Media, 2020, See Leah Palmer's "Foreword to Chapter Seven" pp. 223-228 and Chapter Seven: "Love, Scripture, Suffering and Bioethical Questions", pp. 229-254; and see Adriana Vasquez's Foreword, pp. 84-91, to Chapter Two: "Marriage is a Liturgical Act" of Etheredge's *The Human Person: A Bioethical Word*, 2017.

[122] Cf. Etheredge, *Conception: An Icon of the Beginning*: the following is a profound analysis of the evidence by Profs.

Why is it only a question of arguments based on often difficult philosophical positions when, in reality, there is a wealth of human experience, particularly the experience of the woman and mother?[123]

It is true, however, that there are many and varied injustices in the treatment of women and, in particular, in the abuse of women that leads to pregnancy; however, these problems need addressing independently of the child that may be conceived as a result of this mistreatment, just as children need help in families where the parents are struggling in other ways.

The Witness of Each One of Us

In contrast to the uncertainty that surrounds human conception, there is no doubt that a person who comes to

---

Justo Aznar Lucea and Julio Tudela: Chapter 5: Part II: "The Biological Status of the Early Human Embryo, When Does the Human Being Begin?", pp. 480-507.

[123] Cf. Etheredge, *Conception: An Icon of the Beginning*, for Scriptural citations under the heading: Chapter Two: "The woman's perception of conception: Part V of XII", pp. 158-162 and drawing further on Job up to p. 166.

exist, comes to exist at a certain point in time. In other words, each one of us is an indelible, irrevocable and incontrovertible witness to the three-dimensional fact: firstly, each one of us comes to exist; secondly, we come to exist amidst multiple relationships, beginning with those from whom we received our ordinary human inheritance, ordinarily our parents; and, thirdly, that our life is a gift and, if we are well disposed to recognizing it, we recognize it as a gift. Whatever we may think about the multitude of questions that surround human conception, that we are conceived is certain: that conception, which means 'beginning', is certain. The contrary claim, that we did not come into existence, is clearly contrary to the facts of the union of sperm and egg and the existence of each one of us. Even with respect to the manipulation of human life which involves the multiple injustices of fertilization in a glass dish, discarding unwanted human embryos, freezing of human embryos, experimenting on human embryos and combining different paths to the whole of conception, wherever there is the living human body there is the presence of the person. Otherwise why would husband and wife, from the dawn of time, come together and exclaim, like Eve, 'I have gotten a man with the help of the Lord' (Gn 4: 1)? Otherwise why would a human egg be fertilized in a 'petri dish' in 1977, and born in 1978, make the birth of Louise Brown 'headline news around the world'[124]? Otherwise why would a frozen human embryo returned to the nurturing womb be a human being: a girl called Hannah?[125] Thus there is the corresponding human right of the human life, once conceived, in whatever way he or she is conceived, to the completing nurture of maternal implantation and development.

A Discussion on the Teachings of the Catholic Church and the "Opinion of the Court"

In the course of the "Opinion of the Court", although there was a review of ancient views on human conception, the judge said that it is now the 'official belief of the Catholic Church' to 'recognize the existence of life from the

---

[124] Dr. Elizabeth Rex, "End Word" on p. 596 of Etheredge, *Conception: An Icon of the Beginning*.

[125] Dr. Elizabeth Rex, "End Word": '1998 – Hannah Strege, the world's first adopted frozen embryo is born in San Diego, California on December 31, 1998' on p. 599 of Etheredge, *Conception: An Icon of the Beginning*.

Chapter Seven: An Answer to the Uncertainty   207

moment of conception'[126]. But, having said that, it appears that St. Thomas Aquinas' antiquated biology, among other sources, was used to justify the claim that ordinarily there is not a human being from the first instant of conception [127] because, in the "Opinion of the Court", the judge said: 'We need not resolve the difficult question of when life begins. When those trained in the respective disciplines of medicine, philosophy, and theology are unable to arrive at any consensus, the judiciary, at this point

---

[126] Cf. Pp. 160-161 of pp. 113-178 of ROE v. WADE Syllabus ROE ET AL. v. WADE etc.

[127] Dr. Elizabeth Rex: "End Word": '1973 – On January 22, the U.S. Supreme Court legalizes abortion in Roe v. Wade. The majority decision uses 13th century theology and science to support their erroneous decision about when human life begins, stating: *"Christian theology and canon law came to fix the point of animation at 40 days for a male and 80 days for a female,* a view that persisted until the 19th century…. *Due to continued uncertainty about the precise time when animation occurred, to the lack of any empirical basis for the 40-80 day view, and perhaps to Aquinas' definition of movement"* (footnote 807: Roe v. Wade, 410 U.S. 113 (1973) IV.3, with emphasis and italics added) on p. 611 of Etheredge, *Conception: An Icon of the Beginning.*

in the development of man's knowledge, is not in a position to speculate as to the answer'.[128] Firstly, then, the statement of the judge clearly neglects to consider his own observation that Catholic Teaching, resourced as it is by numerous experts, now teaches 'the existence of life from the moment of conception'. Secondly, the judge showed signs of recognizing the growth of a consensus when he said: 'As one brief amicus discloses, this is a view strongly held by many non-Catholics as well, and by many physicians'[129]. Thirdly, the embryological evidence which would have helped the judge recognize the beginning of human life resulted in the opposite, namely, misleading him to think that there was no clarity about the beginning of human life; and, therefore, as an objection to the view of a first instant of human conception the judge said: 'new embryological data ... purport to indicate that conception is a "process" over time, rather than an event'[130]. In other

---

[128] P. 159 of 113-178 of ROE v. WADE Syllabus ROE ET AL. v. WADE etc.

[129] P. 161 of 113-178 of ROE v. WADE Syllabus ROE ET AL. v. WADE etc.

[130] P. 161 of 113-178 of ROE v. WADE Syllabus ROE ET AL. v. WADE etc.

Chapter Seven: An Answer to the Uncertainty   209

words, there is no inconsistency between conception as a process over time and a first instant of fertilization. For, a process over time has a beginning, namely that there is a first instant that the sperm is enclosed by the closure of what was the egg's open pores and thus the walled embryo is the first stage of a new entity: the nascent human being; this first instant of the human embryo is in contrast to the prior, separate existence, of the active human sperm and the inert human egg. Furthermore, then, the process of fertilization is simply the first stage of human growth from the first instant of fertilization to the fusion of the respective nuclei of what was a sperm and an egg and the expulsion of unrequired chromosomes. Normal development, as even the Warnock Report concluded, proceeds uninterruptedly from the beginning onwards[131]. Fourthly,

---

[131] It is necessary to note that even the Warnock Report recognized that there is a seamless process of development: 'there is no particular part of the developmental process that is more important than another; all are part of a continuous process' (Department of Health and Social Security [UK], *Report of the Committee of Inquiry into Human Fertilisation and Embryology* (London: Her Majesty's Stationery Office, July 1984), para. 11.19, quoted in Catholic Bishops' Joint Committee on

the presence of 'new medical techniques such as menstrual extraction, the "morning-after" pill, implantation of embryos, artificial insemination, and even artificial wombs' seemed to have confused the issue of when life begins because each of these medical techniques required a careful examination of their relationship to the beginning of life and, therefore, should have at least raised a cautionary note about their use as an objection to there being a moment of the beginning of human life. Indeed, as already noted, where the human body lives there is the presence of human personal life – especially bearing in mind the principle that biological development is inherently, as it were, psychological development: being a male

---

Bio-ethical Issues, *Response to the Warnock Report on Human Fertilization and Embryology* (London: Catholic Media Offices, 1984), 13. In response, the Catholic Bishops of Great Britain said that 'our society should resolve to protect the life of the human embryo *precisely from the beginning of its continuous development, ie, from conception (fertilization)*' (Catholic Bishops' Committee, *Response to the Warnock Report*, 13). In other words, the very truth of the seamless process of development requires the welcome of each human life from conception.

or female person is to be a 'psychologically inscribed embryologically-begun human individual'[132].

Finally, then, the following two claims are problematic: that 'We need not resolve the difficult question of when life begins' and 'When those trained in the respective disciplines of medicine, philosophy, and theology are unable to arrive at any consensus, the judiciary, at this point in the development of man's knowledge, is not in a position to speculate as to the answer'[133]. They are problematic because, in effect, the "Opinion of the Court" has speculated about when life begins, calling it 'potential life', and has concluded that the state need not defend the interest of the nascent human life from his or her very beginning. In other words, instead of the court ruling that the question of the beginning of human life required further investigation and withdrawing from making a judicial judgement, the court implicitly held a quasi-philosophical view of there being a 'potential life' and permitted the destruction of a child from his or her beginning.

---

[132] Etheredge, *The Human Person: A Bioethical Word*, p. 311.

[133] P. 159 of 113-178 of ROE v. WADE Syllabus ROE ET AL. v. WADE etc.

A Clarification as Regards the Teaching of St. Thomas Aquinas

As regards the view of St. Thomas Aquinas; he argued, not simply that there are three stages to ordinary human development but that, as a whole, nature intends a man:

> 'St. Thomas embraced such a comprehensive account of Christian and philosophical thought that it is worth considering his understanding of the implication of the delayed ensoulment of a human being; he said: 'foetuses are animal before they are human ... [but] nature, in producing the animal foetus, is aiming at producing a man'[134].

What is more, the philosophical biology on which St. Thomas based his three phase conception of human being was Aristotelian in that Aristotle held that matter was eternal and that, therefore, form (which determines what first matter will be) was always necessary to differentiate

---

[134] *Summa Theologiae*, Methuen, Pt I, Qu 85, art 4, p. 136; but quoted from Etheredge, *Conception: An Icon of the Beginning*, p. 237.

the matter that eternally existed into its various kinds; and, as it was held that human conception was not developed enough to receive a human soul, so it was understood that there was a three phase ensoulment: plant; animal; and then rational ensoulment [135]. But, all the while, St. Thomas held the sophisticated view that nature, nevertheless, 'is aiming at producing a man'; and that, therefore, even modern embryology confirms the view that the fruit of spousal union, ordinarily, in fact and from conception intends a man or a woman.

What seems to be much less well known, is that St. Thomas Aquinas also argued that the conception of Christ was immediate: 'conception of the body [does not precede] ... animation by a human soul ... in Christ'[136]; and, indeed, it is this understanding, explicit in St. Maximus the Confessor, which advances the view that what happened to Christ, notwithstanding the virginal

---

[135] For a more comprehensive discussion of this go to Etheredge, *Scripture: A Unique Word*, Cambridge Scholars Publishing, 2014, pp. 303-306.

[136] *Summa Theologiae*, Methuen, Pt III, Qu 5, art 5, p. 484, footnote 239 of p. 233 of Etheredge, *Conception: An Icon of the Beginning*.

conception, is what happens to us.' Fr. John Saward, drawing on St. Maximus the Confessor, says: 'Apart from the saving novelty of its virginal manner, the conception of Christ is in all respects like ours'[137]. In other words, while there is an argument from St. Thomas concerning the delayed animation of a being from conception by a rational soul – there is a more important argument that Tradition has taken up, namely, that human conception follows, however imperfectly, the conception of Christ: that just as Christ was one in body and soul from conception so we are one in body and soul from conception.

The Contribution of Revelation and Dogma

Drawing 'on St. Maximus, Fr. Saward says that 'if the embryo immediately after fertilization is endowed with only a vegetative soul, then men father plants, not men. But in fact the act of fertilization establishes a human-to-human relationship between father and child; *I am*

---

[137] In footnote 249 of *Conception: An Icon of the Beginning: Redeemer in the Womb*, Ignatius Press: San Francisco, 1993, p. 12, but see also pp. 8-13.

Chapter Seven: An Answer to the Uncertainty

conceived by my father"[138]. Alternatively, what is the living human life which has begun: Is it not an actual human life with the potential of manifesting the whole presence of the human person? Within the very tradition of the Church, then, there is the unique conception of Christ drawing us ever closer to the truth of human conception, one in body and soul from the first instant of human existence; and, in addition, there are scriptural and other dogmatic resources to be drawn upon to elucidate the mystery of human conception. Indeed, notwithstanding the different accounts of human conception and their authors' purposes[139], there is a sense that who is conceived is conceived as a whole; as David says:

---

[138] Etheredge: *Conception: An Icon of the Beginning*, p. 221, quoting from *Redeemer in the Womb*, Ignatius Press: San Francisco, 1993: p. 10, drawing on the Ambigua 2, 42; 1337B-1340B.

[139] In Genesis, for example, there is a very different account to the creation of human being, male and female, to that expressed elsewhere and yet its very difference is open to interpretive explorations, albeit not contradicting natural truths (according to St. Augustine); see "Chapter Two: Scripture and the

'Thy eyes beheld my unformed substance; in thy book were written, every one of them, the days that were formed for me, when as yet there was none of them' Psalm 139: 16; and, indeed, the very unique Hebrew word, *golmi*, communicates an incredible summary sense of 'an unfinished vessel' which is inescapably personal: my unfinished vessel[140].

At the same time there is the mystery of Mary, the Mother of the Lord which has, in the last century, come into great and greater prominence for a variety of reasons but, in truth, to assist us in a timely and providential understanding of both the mystery of the Church, the mystery of salvation in Christ and the mystery of each one of us. This great untapped mystery offers us a singular glance at the moment of human conception, primarily so that we can understand that Christ inherits human flesh free from original sin – but nevertheless it does so in such a way as

---

Beginning of Human Being", *Conception: An Icon of the Beginning*.

[140] Cf. pp. 190-200 of *Conception: An Icon of the Beginning*.

to illuminate the moment characteristic of human conception:

> 'If grace requires the presence of the human soul, then for grace to be effective in the flesh, as it were, as well, then body and soul need to be united. Thus the mystery of the *Immaculate Conception* implies that Mary is one in body and soul (*Gaudium et Spes*, 14) at the instant of their reciprocally coming to exist; indeed, as it says simply in *Lumen Gentium*: 'Enriched from the first instant of her conception with the splendor of an entirely unique holiness, the virgin of Nazareth is hailed by the heralding angel, by divine command, as "full of grace" (cf. Lk. 1: 28 ...)' (56). In other words, while the Church does not explain the 'first instant of conception' – the ultimate 'first instant' is the first instant that the sperm animates the egg and the embryo expresses this through the formation of the embryonic wall'[141].

---

[141] Etheredge, *Mary and Bioethics: An Exploration*, p. 166; and the footnote to this quotation goes to Etheredge, Chapter 12, *Scripture: A Unique Word*, in which the intricate evidence for these claims is discussed.

In a word, then, just as St. Thomas Aquinas argued that we need the help of Revelation to aid our understanding of whether or not there was a beginning to creation so we need the help of Revelation to determine the truth concerning human conception: a truth which confirms and expresses the common understanding of personal experience.

A Variety of Bioethical Declarations

Controversies, as we know, have raged and will continue to rage down the centuries of human history; but, in contrast, we have a number of helps. The Hippocratic Oath states: 'I will not give a woman a pessary to procure abortion'[142]. The Nuremburg Code says: 'No experiment should be conducted where there is an *a priori* reason to believe that death or disabling injury will occur'[143]. The Belmont Report says: 'persons with diminished autonomy

---

[142] Courtesy of Dr. Mary Anne Urlakis, p. 21 of *The Human Person: A Bioethics Word*.

[143] https://media.tghn.org/medialibrary/2011/04/BMJ_No_7070_Volume_313_The_Nuremberg_Code.pdf.

are entitled to protection'[144]. Indeed there are any number of wonderful declarations that seek to draw the truth from human experience, rectify wrongs and establish a way forward for us all; and, indeed, as it has been discussed, it involves the use of technical terminology which, however, has an understandable significance: the reality that each one of us begins: a beginning which entails an inviolability which requires recognition and, where appropriate, the remedies of medical help which are possible and applicable for the benefit of each human subject, whether embryonic or adult. In a word, however, there is an ongoing necessity that there be an explicit recognition of what constitutes both the historical truth of what was intended by specific national legislation and its updating according to a more explicit understanding of a relevant reality, such as human conception; but, also, we are in a new human context that requires a renewed understanding that even specific, national laws, exist in the context of universal truths and rights and that, in the end, these are a part of what will fashion the future for all of us. Thus, it seems, we are already beginning to see the articulation of

---

[144] https://www.hhs.gov/ohrp/regulations-and-policy/belmont-report/read-the-belmont-report/index.html.

principles which are capable of being the basis of national and, possibly, international law: 'it is clear that the principle of respect for human dignity, due to its non-negotiable character and overarching scope, will always play a crucial role in every decision concerning biomedical practice'[145].

Gravitating to a Consensus

Without a beginning there cannot be development and without development there cannot be a precise perception of what began; but, on the basis of what unfolds from a beginning, so what began from that beginning is made manifest. In the words of Pope Francis, collaboratively expressed with the Grand Imam Ahmad Al-Tayyeb, "'In the name of innocent human life that God has forbidden to kill, affirming that whoever kills a person is like one who kills the whole of humanity, and that whoever saves a

---

[145] "Principles of international biolaw: Seeking common ground at the intersection of bioethics and human rights": Roberto Andorno, Bruylant, 2013: "Chapter 1: Principles of International Biomedical law", p. 19 of pp. 13-35: https://www.academia.edu/4063596/Principles_of_international_biolaw.

Chapter Seven: An Answer to the Uncertainty 221

person is like one who saves the whole of humanity …'[146]. How true this is turning out to be when whole countries permit the destruction of the child, not to mention international organisations which promote abortion,

---

[146] Pope Francis, *Fratelli Tutti*: http://www.vatican.va/content/francesco/en/encyclicals/documents/papa-francesco_20201003_enciclica-fratelli-tutti.html#_ftnref262.
In *Conception: An Icon of the Beginning*, there is a Muslim scholar who holds the view of human life from conception, p. 575: 'Germany, however, has gone before the world and enacted the following, 1991 legislation: "Act for the Protection of Embryos" (The Embryo Protection Act). This "word" of law, as it were, has become a world-wide teacher and a noted Muslim bioethicist commented, approvingly, on the German law: Hassan Hathut (1924-2009) 'referred to Germany, which banned all use of human embryos in biomedical research. As for the surplus of fertilized ova in the IVF processes, the law even banned initiating such a surplus …. Hathut concluded that this law goes in line with Islamic ethics (Hathut 1994, 175)' (citation on the Muslim scholar from: '"Islam, Paternity, and the Beginning of Life": "The Beginning of Human Life: Islamic Bioethical Perspectives" with Mohammed Ghaly, (*Zygon: Journal of Religion and Science*, vol. 47, No. 1, (March 2012), pp. 175-213: https://core.ac.uk/download/pdf/43497555.pdf, p. 207.'

pharmaceutical and IVF companies which have no regard for the truth that human life begins at the first instant of conception and that each person has a right to an integrally human identity and his or her completing human development. By contrast, however, there is a growing international recognition of these human rights which are beginning to be reflected in what is called the "Geneva Consensus Declaration": the Center for Family and Human Rights 'has worked for 24 years toward the declaration made by the Trump administration today together with a coalition of 32 UN Member States. There is no international right to abortion. There is no international obligation to fund abortion. The United Nations has no business interfering in sovereign decisions when it comes to protecting life in the womb' (October 22, 2020)[147].

---

[147] First paragraph of "Statement of Austin Ruse, President of C-Fam, on the signing of the Geneva Consensus Declaration" (October 22, 2020): https://email.opusfidelis.com/t/ViewEmail/j/92252BD7C1C8076B2540EF23F30FEDED/80B27ACA6DF63276F7E8006BBCB98688; cf. also: Austin Ruse, 23 October, 2020, "Governments Launch Pro-Life Declaration at United Nations: https://c-fam.org/friday_fax/governments-launch-pro-life-declaration-at-united-nations/.

This is not a naïve proposal – but would clearly require patient, persistent and prudent international cooperation[148]. As Roberto Andorno has said: 'Global challenges raised by biomedical advances require global responses'[149].

---

[148] A relevant essay by Teresa Etheredge, and an article by Grace Etheredge, have helped me to think through these questions: Teresa Etheredge: "Why is the criminological imagination important to the future of criminology? Describe a current issue in criminal justice and explain how the criminological imagination could help us to understand it"; and Grace Etheredge: "The impact of public international law on UK courts".

[149] "Biomedicine and international human rights law: in search of a global consensus": "Abstract": p. 960 of the "Bulletin of the World Health Organization 2002, 80 (12)".

PART V

UNFOLDING A POST-ROE WORLD

Comprising Chapters Eight, Nine and Ten

In this book there have been many discussions, whether of law, imagery, gardening, the word of God or theology; and, whether implicitly or explicitly, there are both immense pastoral works and philosophical questions that continue to be addressed.

In the final chapters of this book, we begin to see how the forthcoming judgement, Dobbs v Jackson, begins to open, anew, a variety of concerns basic to justice, truth, and human compassion. As we progress through these chapters, there emerges a clearer sense of the "light" and the "dark" of the times in which we live and the imperative, day by day, of reaching the embracing truth of our nature, from conception onwards, that nobody who belongs to the human race is excluded and everyone who is included, is justly addressed in their full and common humanity.

At the same time, it is necessary to make a distinction that, while it is often alluded to, has not been spelled out: the distinction between what is actual and potential.

Distinguishing Actual and Potential Human Life

"You cannot become what you are not". How coherent is the idea of 'potential [human] life'? A potential is a possibility; it is not an actuality. Therefore, prior to fertilization, there is a potential human life, a possibility of a human life being begun; however, once a real conception has occurred, there is not a 'potential life' - there is an actual human life full of potential. A human life, by definition, then, has to exist to be full of potential; and, if a human life exists, then it cannot be full of potential if it is not already wholly what it is.

In other words, a human being cannot become a human being if he or she is not a human being already. Therefore, there is no watershed that transforms the beginning of a human life into a human being; rather, there is an ongoing manifestation of what is present from the beginning. If you light a fire made of wood, you begin the transformation of the fuel into ashes: the actual flame has the potential to burn up the whole fuel; however, if you have once lit the fire and then put it out, there is no possibility of the rest of the fuel being burnt.

In other words, if you exist, if you are actual, then who you are is increasingly made manifest; however, if what you are is "put out" no amount of imagining the potential that had been there will bring it back. If you put the fire out - you also put out the potential of it burning the rest of the fuel!

CHAPTER EIGHT

ROE V WADE:
ONGOING ARGUMENTS OF BENEFIT TO US ALL

As the discussion continues about the relationship between the American Constitution, those who object to abortion and those who advocate abortion, there is emerging a certain amount of clarity around a number of central arguments which, indeed, have a world-wide relevance to the debate concerning the legal recognition of the right to life of the living.

A general point, however, before moving on is the circumstantial situation of women in society and, depending on the immediate culture, there may be more or less support for even a welcome pregnancy, never mind one that is ether unexpected or even forced upon the woman. In other words, there are always wider issues than that of the immediate question of the right to life; and, in justice, none of them can be neglected to the detriment of the whole provision of help to family life[150]. In particular, and admittedly in the specific context of America, there are many significant changes in society, all of which need their own analysis and synthesis, but one of them is the rise of the number of men and women in prison: between a steady population of 10,000 at the turn of the century it has risen to 1, 414, 162, in 2018[151]. In other words, during

---

[150] It is a gross oversimplification to identify the Catholic Church's response to abortion and the whole range of human rights violations with one or two documents of the Church when there are so many religious orders, priestly works and lay responses that, in justice, need to be recognized; cf. Beattie, "Catholicism, Choice and Consciousness: A Feminist Theological Perspective on Abortion", p. 57.

[151] Unpublished text supplied by the author and with permission to cite: Soozi Scheller, (with Dr. David S. Crawford, Law, Family, and the Person), "Unfruitful Law: The Woman as a Revealer of the Heart of the Culture", 23rd April, 2021, p. 6: 'Between 1925 and early 1960s, the population in the U.S. state and federal prisons had remained at about 10,000 men and women. However, by 1964, this figure had increased to 20,000 and further by 40,000 in 1984. With each subsequent decade, the increase became more rapid, so that in 2018 the number of incarcerated individuals reached 1,414,162, which is the highest

Chapter Eight: Roe v Wade: Ongoing Arguments 233

the 60s there began a trend which has only increased, namely, that just as natural law has become obscured and marginalised so has the social transgression of the law of the land increased[152]. One wonders, then, more widely, how socially disorientating is the loss of reason's guide to moral action[153].

---

per capita rate in the world' (footnote in the original to "Trends in US Corrections" etc).

[152] E.g. The rise of hormonal contraception (which obscures the key woman's health indicator of ovulation), abortion and confusion about male and female identities); my inference but on the basis of some of Soozi Scheller's essay, "Unfruitful Law" etc. Michal Paszkiewicz comments, in an email, 29/09/2022: "I imagine it is probably very likely an effect, but I'm not convinced the US presents a strong enough case... The US actually decreased incarceration rates from 1960 to 1970. The UK had a less abrupt growth in incarcerations from 1940 onwards and only shot up faster in the early 90s. Incarceration rates can also change not due to changing numbers of transgressions, but stricter rules, or stricter enforcement." There is, clearly, the possibility of a wider discussion of this theme.

[153] Cf. The forthcoming work, *Human Nature: Moral Norm*, by Francis Etheredge and from En Route Books and Media, maybe late 2022.

There are the following seven sections to this postscript: Justice Beyond a Change of Justices (i); Viability is for Life (ii); Choice, Burdens and their Alleviation (iii); Bodily Integrity, Liberty, Equality and the Constitution (iv); Brain Death and Abortion (v); Abortion and the Advancement of Women (vi); True Justice is Irreversible (vii).

Section i: Justice Beyond a Change of Justices. If whether or not a court decides to grant a legal right to abortion depends on who has appointed the judges[154], then there is a problem with the criteria by which a legal decision is made; and, as such, it turns the question as to who lives and dies into an absolute lottery – for all who are the legally unrepresented subjects of a decision to allow any kind of deliberate abortion, experimentation or otherwise infringement of the right to protected continuing human development. In other words, the very nature of what makes these variable decisions possible is already demonstrating to the nation and the world that politicians

---

[154] Cf. Mark Sherman, November 29, 2021, AP News: "Supreme Court set to take up all-or-nothing abortion fight".

and 'the Court … [are] playing politics'[155]. But the point, surely, is not whether or not the Court or others are 'playing politics' but that there is a public duty for a court to determine what is just and just for all; and, therefore, both the Court and politicians cannot renounce their responsibility, before God and their nation, to formulate just laws. In other words, democracy is not an end in itself – it needs the structure of truth to assist it; otherwise, one court is exchanged for another, or one judge for another, without an advance in the relationship of truth to justice: the rights that are due to each one of us, born or preborn.

While there are those who think that an outcome of the current debate could be a 'limited "right to abortion"'[156], Justice Clarence Thomas says that there is not a 'shred of support' for a right to abortion in the American Constitu-

---

[155] Helen Alvaré, December 1st, 2021 : "Supreme Court's 'Dobbs v. Jackson' Oral Argument" etc: https://www.ncregister.com/commentaries/supreme-court-s-dobbs-v-jackson-oral-arguments-promising-for-pro-life-cause.

[156] David *Bjornstrom, Esq, "Dobbs Case Exposes"* etc: https://personhood.org/2021/12/08/dobbs-case-exposes-moral-bankruptcy-of-abortion-advocates/.

tion[157]. In view, then, of the clear statement of the 14th Amendment, that no State shall 'deprive any person of life, liberty, or property, without due process of law' – it is possible that a review of what the Constitution says will have to address the obvious contradiction between a supposed 'right to abortion' and the prohibition against depriving anyone of life. At the same time, 'Some of the briefs ask the Court to let each state decide abortion law for itself'[158] – but again this does not go beyond the variability of who is elected when[159], whereas for those who

---

[157] Jamie Ehrlich, July 9th, 2020: https://edition.cnn.com/2020/06/29/politics/clarence-thomas-abortion-dissent/index.html.

[158] David *Bjornstrom, Esq*, *"Life or Death"* etc: https://personhood.org/2021/10/28/pro-life-and-pro-abortion-arguments-in-the-dobbs-case/.; *cf.* also Steven Ertelt, *"Supreme Courts Lets"* etc: https://www.lifenews.com/2021/12/10/supreme-court-lets-texas-abortion-ban-keep-saving-babies-dismisses-joe-bidens-lawsuit-against-it/.

[159] Cf. Mary Ziegler, October 3rd, 2022, "When Fetal Rights Are More Important Than Democracy": https://www.theatlantic.com/ideas/archive/2022/10/anti-abortion-movement-roe-v-wade-democracy/671583/.

are deliberately aborted, it is an irrevocable outcome for them.

But whatever comes of the current debate in America, this book has established the right to legal representation from conception on the basis that human rights are integral to human relationships; and, therefore, a human being is by definition a human being-in-relationship and, as such, entitled to the rights embodied in being human. Thus we need principles and practices that are not based on "Legal Positivism", the doctrine that a particular legal decision is *per se* right and irreformable just because it is embodied in a specific law. We need, rather, a principle of justice expressed in a law for the sake of all: non-discriminatory; irreformable; and adequate to all the needs of the child.

Section ii: Viability is for Life. It is reported that 'viability' is roughly understood to be equivalent to 'when a fetus can survive outside the womb'[160]. Thus there are a number of problems with the term 'viability'. Whether or not a child can survive being prematurely born or

---

[160] Mark Sherman, "Supreme Court set to take up all-or-nothing abortion fight".

delivered is almost entirely dependent on the level of available technical equipment, expertise and the skill to manage the maturation of breathing and organ development. In other words, 'viability' is here understood to be dependent, or contingent upon, factors which have nothing to do with the inherent right of living in relationship to other human beings; indeed, just as a dependant has a right to be fed, clothed and educated, so the extreme dependency of a baby is even more a reason to respect the right to succour and support in his or her struggle to live.

'Viability', however, has a natural meaning of 'capable of living'[161]; indeed, as applied to the sperm and egg, viability means that 'pregnancy depends on the viability of the sperm and egg'[162]. In other words, although viability tends to be used as a legal term, referring to the life of a child being viable or capable of life outside of the

---

[161] Cf. "Viable": https://www.merriam-webster.com/dictionary/viable; and cf. "Fetal Viability": https://en.wikipedia.org/wiki/Fetal_viability.

[162] "Viability": https://www.google.com/search?q=viability+definition&rlz=1C1SQJL_enGB840GB840&oq=Viability+def&aqs=chrome.0.0i512j69i57j0i512l8.7340j0j7&sourceid=chrome&ie=UTF-8.

womb[163], being viable can clearly refer to the life of the child from conception and his or her ongoing development in the womb. Being 'capable of life', then, really expresses the self-evident fact that being alive entails the capability of being alive: that being capable of life is an inherent and intrinsic property of the entity, whether it be sperm, egg or fertilized human embryo. It is almost as if the term, viability, could apply to the idea of *being alive in such a way as to be able to continue to live.* Now, that being the case, just as there are external factors that assist the premature birth of a child to survive being premature, so the natural environment of the human embryo is the mother's womb: the place where, being alive in such a way as to be able to continue to live, indicates an inherent ability to live that is, at the same time, sustained by the natural or assisted environment of the child. In other words, 'viability' really refers to the existence of a child being alive of itself and, by definition, is alive both independently of the mother and of any assistance afforded the child once born. In other words, it is not that the child does not need the mother's womb to develop or that the child does not need

---

163 "Fetal Viability": https://en.wikipedia.org/wiki/Fetal_viability.

specialist help if born prematurely; rather, viability refers to the fact of the life of the child is in itself independent of both mother and any subsequent support. Thus assisting what is already living is rooted, as it were, in the very fact of the human embryo being alive of itself. 'Viability', then, is evidenced in the very moment of conception, as we have seen, from which the embryological development proceeds of itself, *uninterruptedly,* from the first instant of fertilization until natural death.

Section iii: Choice, Burdens and their Alleviation. The problems of the burden of pregnancy and the burdens of parenting are not identical and are equally in need of addressing[164]; but, in either case, there is the question of what choice has already been exercised and, in the event of pregnancy, the difference it makes of another life like that of ours.

With respect to the woman's choice

---

[164] *Micaiah Bilger* | Dec 1 :
2021: https://www.lifenews.com/2021/12/01/justice-amy-coney-barrett-destroys-pro-abortion-argument-that-abortion-bans-force-women-to-be-pregnant/.

Chapter Eight: Roe v Wade: Ongoing Arguments

Justice Roberts says: "…if you think that the issue is one of choice, that women should have a choice to terminate their pregnancy, that supposes that there is a point at which they've had the fair choice, opportunity to [choose], and why would 15 weeks be an inappropriate line?' [165]

Clearly there is a variety of problems here: A woman may not have chosen pregnancy, either because of rape or a failed contraceptive or she has simply changed her mind, on becoming pregnant, and decided not to go through with being pregnant; but, even so, the question is no longer that of a personal choice, as to whether to take the risk of contraceptives, or to become pregnant, or to end the life of a child that has come to exist, precisely because there is another human life that is implicated in the decision of whether or not to have a deliberate abortion. Who, as I say, represents the right of a child at risk if not an impartial authority whose task it is to safeguard the rights of

---

[165] Jonah McKeown, December 2, 2021, « 'Dobbs v. Jackson'" etc: https://www.ncregister.com/cna/dobbs-v-jackson-what-did-roberts-kavanaugh-and-barrett-say.

all human beings, namely a court? In any other conflict of interest the child would be represented in court – Why not now, when, there is a risk to the child's very life? As regards the concept of 'fair choice' – how does that apply to the child? In the end, neither 'fair choice' nor 'viability' are relevant if the court's role is to protect all innocent human life[166].

With respect to the claim that there is a burden of pregnancy, it has to be admitted that there are risks to the mother which, while they can be alleviated, cannot be totally excluded. On the one hand, there has been an appreciation of mothers from ancient times; and, indeed, Tobias says to his son Tobit:

> 'Honor her all the days of your life; do what is pleasing to her, and do not grieve her. Remember, my son, that she faced many dangers for you while you were yet unborn' (Tobit, 4: 3-4). Let us not forget, either, that the Commandment of the Lord carries a promise: 'Honor

---

[166] Lauretta Brown, « After oral arguments" etc: https://www.ncregister.com/cna/after-oral-arguments-in-landmark-dobbs-v-jackson-abortion-case-experts-say-roe-s-days-are-numbered.

your father and your mother, that your days may be long in the land which the Lord God gives you' (Ex 20: 12).

On the other hand, it is possible that gratitude for childbearing has diminished in a culture that often blames a mother for introducing another carbon footprint into the world.

However, the question is whether the risk to the mother is "equivalent" to a "right to abortion". The risk to the mother of continuing the pregnancy is not equivalent to the risk to the child of the mother discontinuing the pregnancy; for, in reality, the child is at greater risk of dying, especially if deliberately aborted, both because of being taken out of his or her natural environment and because of the means by which the child is removed. It is also true that there are many effects on women of what they do, ranging from the psycho-physical to the spiritual; even the so-called home abortion pill has a rising record of ill effects:

'As you may have seen in the media, a study [in the UK] has revealed that over 10,000 women have needed hospital treatment following the use of medical abortion pills since March 2020'[167].

In other words, there is a concrete burden on a woman of abortion itself. Similarly, any use of contraceptive hormones, never mind any other unnatural devices used to prevent or end a pregnancy, has an impact on the woman. Indeed, one of the most tragic effects of hormonal contraception, beyond the possible death of a child, is that it suppresses ovulation which is, apparently, a reliable indication of a woman's health[168].

---

[167] Cf. Email: Right To Life UK info@righttolife.org.uk, 06/12/2021: https://percuity.files.wordpress.com/2021/10/foima-treatment-failure-211027.pdf; and cf. David Maddox, "Abortion Pill Horror": https://www.express.co.uk/lifestyle/health/1527888/Abortion-pill-diy-nhs-warning.

[168] Cf. Marguerite R. Duane and Erin Adams, "The State of Fertility Awareness Based Method Education for Medical Professionals," in *Humanae Vitae, 50 Years Later: Embracing God's Vision for Marriage, Love, and Life*, ed. Theresa Notare (Washington, DC: The Catholic University of America Press, 2019), p. 206: ovulation is monitored "since the presence of ovulation is a sign of good health" (and see footnote 40: Pilar Vigil

As regards the burden of parenting, again it is realistic to acknowledge that there are all kinds of difficulties for parents, single or married, to address; however, we do so in the context of many formative influences, whether of our own parents, friends or the contribution of Church groups.

'We are all sinners and we're all selfish in some ways or others. Parenting reveals those things in a way that can be so painful[,] so it makes sense that we all hate parenting sometimes. But we can't stay in that spot and hope it gets better on its' own!

We really recommend examining what's underneath that feeling of hating parenting. We also recommend looking at your family culture.

Family life is difficult and messy but it doesn't have to be a constant survival mode. When we intentionally focus on community within our homes, and on

---

et al, *The Linacre Quarterly*, no. 4 2012); and Cf. Francis Etheredge, "A Touch of Experience: Where Are You?": https://www.hprweb.com/author/francis-etheredge/.

positive things like joy, taking delight in our children and having fun with them, the atmosphere in our homes can really change'[169].

Having ten children in ten years, two of whom were early miscarriages, it is clear that a variety of help is needed. In the end, however, there are other considerations, namely the rise of Western infertility and the decline in the birth-rate; indeed, these facts raise questions: Why are Western societies subject to these problems?

'Developed countries have experienced unprecedented declines in fertility rates over the last help of the 20th century with a prevalence of subfertility close to 10%, mostly owing to genital track infections'[170]. But then there are other problems in China and India, namely 'the use of other technologies (ultrasonic screening

---

[169] Mike and Alicia Hernon, September 8, 2022, "Help! I Hate Parenting My Kids (Sometimes): https://www.catholicweekly.com.au/help-i-hate-parenting-my-kids-sometimes/.

[170] Anastasiadou and Joep Geraedts, *et al*, "The interface between assisted reproductive technologies and genetics' etc. pp. 620-621 and 622.

and abortion) has lead to sex disparities in China and India'[171].

At the same time, however, there is the provision for those who, for whatever reason, give up their child. Thus one of the main provisions in American law, and there are probably equivalents in other countries and cultures: are the 'safe haven laws, which allow mothers to relinquish their newborns to authorities without fear of repercussions, as an alternative to abortion and the burdens of motherhood'[172].

One final point, as regards the burdens of motherhood and parenting, is the implication of the following claim: 'Abortion advocates argue that legal abortion is necessary to give women equal rights because men do not get pregnant'[173]. Women's equal rights do not, however, depend

---

[171] *Ibid*, p. 621.

[172] *Micaiah Bilger* | Dec 1 : "Justice Amy Coney Barrett". 2021: https://www.lifenews.com/2021/12/01/justice-amy-coney-barrett etc.

[173] David *Bjornstrom, Esq, "Life or Death" etc:* https://personhood.org/2021/10/28/pro-life-and-pro-abortion-arguments-in-the-dobbs-case/.

on any denial of differences between men and women; rather, women's rights are precisely that: the rights of women as women. Rather, there is a woman's right to be helped with the burdens of motherhood and parenting, precisely because these are of their nature activities that entail a husband or a man's involvement. In other words, even in the case of artificial insemination or other practices, the involvement of others implies the woman's right to help with the outcome of becoming pregnant. Furthermore, if there is an act of injustice against the woman, how much more does she have a right to help? But, in the end, there is the injustice to the child, once conceived, if an abortion is proposed; and, therefore, the right of the child to assistance is an inalienable right: a right which cannot be disregarded without perpetrating a radical and irreversible injustice to the child.

Section iv: Bodily Integrity, Liberty, Equality and the Constitution. Although these points pertain to a specific case of American law, the features which are claimed to be relevant to the woman all pertain to the child too: 'bodily integrity, liberty and equality'[174]. Once the sperm has entered the open orifice of the ovum, there is a new entity signified by a unilateral closing of all the external openings in what is now the human embryo. In other words, the first instant of fertilization establishes the bodily integrity of the human embryo; and, by integrity, is meant the foundational wholeness of the new being, albeit it will differentiate into placenta and child in due course. Nevertheless, as the placenta is a necessary, albeit temporary organ regulating the nutrition and waste product requirements of the baby, it falls within the terms of bodily integrity. 'Liberty' pertains to the freedom from the life-threatening actions of others as indeed it pertains to the full development of the individual and his or her talents; for, without bodily existence, there is no possibility of taking advantage of life's opportunities as and when the child matures. 'Equality', too, is an expression of the human identity and dignity which is equal to that of the mother; and, dependent as the child is, this dependence does not

---

[174] Steven Ertelt, 1st December, 2021: "Justice Clarence Thomas". https://www.lifenews.com/2021/12/01/justice-clarence-thomas-makes-it-clear-theres-no-right-to-abortion-in-the-constitution/.

detract from the radical equality of both mother and child being equally in receipt of the gift of life.

Further, there is the question, in the case of the American Constitution, of whether or not 'the Constitution is neutral on the question of abortion'[175]. But if, as was argued earlier, the Constitution was written in a time that presupposed the right to life and the presupposition that citizenship was established at birth, which again presupposes the right to life of the child-to-become-citizen, then this legal frame of reference implies both a right to life and the practical necessity of determining citizenship at the time of birth. At the same time, if historically it is clear that the Constitution is not neutral on abortion then it is also clear that either the American people have a referendum or make the Constitution, from this point on, neutral on abortion. However, neither a referendum nor making the Constitution neutral on abortion addresses the underlying question of the inherent right to life of an innocent human being, otherwise unrepresented legally and whose life is irreversibly endangered by an arbitrary decision of

---

[175] Justice Kavanagh : « Dobbs v Jackson" etc.: https://www.ncregister.com/cna/dobbs-v-jackson-what-did-roberts-kavanaugh-and-barrett-say.

Chapter Eight: Roe v Wade: Ongoing Arguments 251

viability outside of the womb – when viability includes the very place where the child is naturally viable, namely the womb of the woman.

Section v: Brain Death[176] and Abortion. Is it true that 'unborn babies are like brain dead people who lack the consciousness to feel pain'[177]? On the one hand, it is recognized that a baby is seen 'recoiling when being poked or touched'[178]; and, indeed, the skin of a human being is a natural expression of his or her psycho-physical integrity. But it is also true to say that the life of a baby expresses the progressive nature of human development and, if the child lost a limb, it would not cease to be a child and would continue to develop. If pain censors are characteristic of

---

[176] Cf. Also *Reaching for the Resurrection: A Pastoral Bioethics*: https://enroutebooksandmedia.com/reachingfortheresurrection/.

[177] Micaiah Bilger | Dec 2, 2021: « Doctor Slams Sotomayer ». https://www.lifenews.com/2021/12/02/doctor-slams-sotomayor-to-compare-an-unborn-child-to-a-brain-dead-person-is-wholly-ignorant/.

[178] Micaiah Bilger | Dec 2, 2021: « Doctor Slams Sotomayer » etc.

our skin, then their presence may well be both evident and active as a part of the normal reflex cycle of the nervous system; and, as such, pain will be present at the start of that reflex process. Thus a child losing a limb in the womb would undoubtedly be painful whether or not consciousness is explicitly "experienced". On the other hand, brain death is a controversial condition because it is often if not wholly linked to obtaining organs for transplants and the person's response to stimuli may simply be a sign that he or she is alive – albeit heavily sedated. Furthermore, those engaged in promoting death by assistance freely acknowledge the centrality of stopping the heart as a foundational criterion for death[179]. Similarly, those seeking to ensure the death of the child in the womb use the drug

---

[179] Cf. the forthcoming from *The Catholic Medical Quarterly*, UK, a three part article, "Loneliness, Euthanasia and the Wholeness of Human Personhood" which, among it many sources, quotes: "Doctors seek life-ending drugs that smooth the way for the terminally ill" by Lisa M. Krieger" etc: https://medicalxpress.com/news/2020-09-doctors-life-ending-drugs-smooth-terminally.html; and cf. also "Is an assisted death 'quick and painless'?" by Michael Cook: https://alexschadenberg.blogspot.com/2021/11/is-assisted-death-quick-and-painless.html.

Chapter Eight: Roe v Wade: Ongoing Arguments 253

used to execute prisoners by injecting it directly into the child's heart[180]. In other words, while a dog is for life and would not be treated like this at all without social outrage, a child is for eternal life and suffers immeasurably the rejection and destruction of an often inhospitable un-welcome.

Irrespective, however, of whether or not and how much a child experiences pain; his or her death is an irreversible event; and, therefore, an act of discrimination[181] against the ongoing life and development of that child.

Section vi: On Abortion and the Advancement of Women. The argument that a child is an obstacle to the advancement of women: Do 'women need to abort their

---

[180] Cf. Beattie, "Catholicism, Choice and Consciousness: A Feminist Theological Perspective on Abortion", p. 64.

[181] See the tendency in the UK to recognize the justice of providing for children with specials needs, in this case owing to Down's Syndrome: https://gript.ie/a-huge-achievement-groundbreaking-uk-down-syndrome-bill-passes-second-stage-in-parliament/?mc_cid=e0a9ef2ad0&mc_eid=47b31ab45d.

unborn babies to succeed'[182]? Justice Amy Barrett, a mother of seven children, and indeed many other working mothers, are clearly evidence that motherhood and work and, more widely, fulfilment, are not incompatible. According to St. Edith Stein, the very opposite is the case in that the natural tendency of the woman to be sensitive to the whole of human personhood is precisely the gift to be preserved and contributed, whether domestically or internationally. St. Edith wrote:

> 'Indeed, no woman is only "woman"; every one has her individual gifts just as well as a man, and so is capable of professional work of one sort or another, whether it be artistic, scholarly, technical or any other. Theoretically this individual talent may extend to any sphere, even to those somewhat outside women's scope' .... 'When working out laws and decrees a man might perhaps aim at the most perfect legal form, with little regard to concrete situations; whereas a woman who remains faithful to her nature even in parliament or in

---

[182] Cf/ *Micaiah Bilger* | Dec 1 : "Justice Amy Coney Barrett".

the administrative services, will keep the concrete end in view and adapt the means accordingly'[183].

Section vii: True Justice is Irreversible. There is the argument that it is better not to reverse the law because it 'introduces uncertainty into the law' or, when people are regularly relying on it, to change the law becomes 'disruptive'[184] and therefore, in both cases, the law ceases to be a stable contributor to the necessary order that enables a society to function coherently. Should, then, the abolition

---

[183] From Edith Stein's "Essays on Woman" compiled from lectures given before she entered Carmel. This excerpt is here with permission and is translated from the German by Freda Mary Oben, Ph.D. Copyright© 1987, 1996 Washington Province of Discalced Carmelites ICS Publications 2131 Lincoln Road, N.E. Washington, DC 20002-1199 U.S.A., quoted by Dr. Ronda Chervin, on pp. 65 and 75 of an unfinished manuscript supplied by the author, R. Chervin, of *The Battle for the 20th Century Mind* (Forthcoming from St. Luis, MO: En Route Books and Media, 2022).

[184] John McGuirk, 3/12/2021, "Is the Supreme Court" etc: https://gript.ie/is-the-us-supreme-court-really-about-to-ditch-roe-v-wade/?mc_cid=e0a9ef2ad0&mc_eid=47b31ab45d

of slavery never have taken place? Should allowing the "common man and woman" to vote not have been allowed? Should there be no correction of an unjust law and, therefore, no progress over time of the law expressing, more fully, the full recognition of justice before the law for all?

However, by contrast, if laws are not corrected when recognized to be erroneous then injustice is perpetuated and the rule of law, by implication, called into contempt. Furthermore, justice being the natural objective of the law, any progress in the enactment of just laws or the correction of unjust ones, establishes hope for both these objectives in the future and, in the present, objectively improved conditions for those directly benefitting from the improvement in the law. Finally, the law is a teacher and, as such, has an obligation to teach what can be learnt concerning the right regulation of society for the benefit of all its members.

In the end, however, is it the place of the court to rule on the nature of human life or, rather, to recognize it as a foundational presupposition of justice for all?[185] Clearly, the whole reason that there is a persistent objection to the

---

[185] John McGuirk, 3/12/2021, "Is the Supreme Court" etc.

legalization of abortion is that it enacts an action against the life of human beings: the giving of which is non-discriminatory and is given by God unconditionally. Indeed, it could be argued, a just law is that which is non-discriminatory, unconditional and irrevocable in its recognition of the right to life of the innocent. In other words, human life needs the help of the protection of international law.

Chapter Nine

A New Beginning

No debate, as long standing and as deep as this one, stands still; and, as in general, there are always new sources, whether ancient or modern, it is always a matter of updating the discussion; however, in the times in which we live, there is the publication of the official judgment of the Supreme Court and, therefore, the actual text to discuss. Furthermore, continual reading raises, too, new aspects of the contemporary situation, both scholarly and in the case of personal testimony which, as we shall see, is capable of influencing presidential policy because of it being evidence of the truth that a human embryo is a human person: a human being-in-relationship. On the one hand, then, there is a growing understanding of the Constitution of the United States and, in addition, how this helps us to understand the capacity of law, more widely, to contribute to a just society; and, on the other hand, there are the numerous ways that the reality of embryo adoption brings to the fore the reality of the embryonic child: not a "blob of cells" – but the nascent beginning of human personhood that requires a place in our hearts just as he or she depends upon a place under the maternal heart to grow and develop.

Moreover, what is striking is that 'Although not well determined, it is known is that many healthy children have been born from supposedly poor quality embryos'[186].

The second edition of this book was prompted, then, by the publication of the American Supreme Court's Opinion, "Dobbs v Jackson"[187], a book by John

---

[186] J. Aznar-Lucea, *et al*, pp. 110-111 of "Frozen Embryo Adoption", *THERAPEÍA*, 8 [Julio 2016], 103-120, ISSN: 1889-6111: https://www.academia.edu/30973678/FROZEN_EMBRYO_ADOPTION?email_work_card=view-paper.

[187] Dobbs, State Health Officer of the Mississippi Department of Health, et al. v. Jackson Women's Health Organization et al.: https://www.supremecourt.gov/opinions/21pdf/19-1392_6j37.pdf.

Strege[188] and two ITEST[189] webinars entitled "A Post-Roe World"[190] and "Bioethics and Law: Understanding the Nexus: Truth and Meaning in Constitutional Jurisprudence"[191].

The Opinion of the Supreme Court

What is clear, then, from "The Opinion of the Court" is that there is no implicit endorsement of a so called right to abortion in the American Constitution, however it is to be found and defined; and, therefore, the Supreme Court

---

[188] John Strege, A Snowflake named Hannah: Ethics, Faith, and the First Adoption of a Frozen Embryo, Kregel Publications, 2020.

[189] ITEST stands for "Institute for Theological Encounter with Science and Technology".

[190] Go to: "A Post-Roe World with Dr. Pat Castle and Kiki Latimer": https://www.youtube.com/watch?v=ms4NEi56ZPY.

[191] WCAT TV presents . . . ITEST Webinar with Fr. Thomas Davis on Bioethics and Law: Understanding the Nexus: Truth and Meaning in Constitutional Jurisprudence: Bioethics, Babies & Bromides": https://youtu.be/928snYsmT3k.

has returned the discussion to each state's legislative body, that each state may determine whether or not to permit abortion, as at the time of the 14th Amendment, 'by citizens trying to persuade one another and then voting' (p. 6 of the Opinion of the Court). Thus the main argument of the "Opinion of the Court" is to show that the stability of the law is best served by overturning a law which has no foundation in the American Constitution[192]. The following three quotations, then, are not an exhaustive account of the "Opinion of the Court" but show the general orientation of the Court's judgement.

---

[192] The Syllabus is not a part of the « Opinion of the Court » but gives an overview of it, pp. 1-2: ': The Constitution does not confer a right to abortion; Roe and Casey are overruled; and the authority to regulate abortion is returned to the people and their elected representatives. Pp. 8–79. (a) The critical question is whether the Constitution, properly understood, confers a right to obtain an abortion. Casey's controlling opinion skipped over that question and reaffirmed Roe solely on the basis of stare decisis. A proper application of stare decisis, however, requires an assessment of the strength of the grounds on which Roe was based. The Court therefore turns to the question that the Casey plurality did not consider. Pp. 8–32.'

'The right to abortion does not fall within this category [of—the Due Process Clause of the Fourteenth Amendment]. Until the latter part of the 20th century, such a right was entirely unknown in American law. Indeed, when the Fourteenth Amendment was adopted, three quarters of the States made abortion a crime at all stages of pregnancy. The abortion right is also critically different from any other right that this Court has held to fall within the Fourteenth Amendment's protection of "liberty ... [as] abortion is fundamentally different, as both Roe and Casey acknowledged, because it destroys what those decisions called "fetal life" and what the law now before us describes as an "unborn human being."[Footnote 13: Miss. Code Ann. §41–41–191(4)(b) (2018)][193].

'It is time to heed the Constitution and return the issue of abortion to the people's elected representatives. "The permissibility of abortion, and the limitations upon it, are to be resolved like most important questions in our democracy: by citizens trying to persuade one another and then voting." Casey, 505 U. S., at 979

---

[193] Dobbs v Jackson, p. 5.

(Scalia, J., concurring in judgment in part and dissenting in part). That is what the Constitution and the rule of law demand'[194].

'Our opinion is not based on any view about if and when prenatal life is entitled to any of the rights enjoyed after birth. The dissent, by contrast, would impose on the people a particular theory about when the rights of personhood begin. According to the dissent, the Constitution requires the States to regard a fetus as lacking even the most basic human right—to live—at least until an arbitrary point in a pregnancy has passed. Nothing in the Constitution or in our Nation's legal traditions authorizes the Court to adopt that "'theory of life.'" Post, at 8'[195].

So, it could be argued, we need an account of 'fetal life', of an 'unborn human being' that will, by its truth and simplicity, attract universal acclaim and juridical approval, throughout the world. Is this an unfounded dream or a natural project: that human beings want to identify the

---

[194] Dobbs v Jackson, p. 6.

[195] Dobbs v Jackson, pp. 38-39.

distinguishing characteristics of their own species in order to offer the future of mankind an irrefutable reference point for the identity of the human race? And, just as court seeks stability in its understanding of a law or a constitution, so human beings seek the stability of truth to transcend the tendency of the human heart to make arguments that justify any means to any goal. Truth, then, is not just for the sake of a specific group of people; it is, rather, an indispensable ingredient of peace.

A Book by John Strege: Hannah: The First Child Adopted as a Frozen Embryo[196]

In one of the leading examples of what becomes of each of us, from conception onwards, there is the life of Hannah Strege; frozen, along with nineteen others, all her siblings, in an *in-vitro* cold storage container[197]. In the end only one child, Hannah, survived the process of thawing, transferring and implanting in the womb of the adopting

---

[196] John Strege, *A Snowflake named Hannah: Ethics, Faith, and the First Adoption of a Frozen Embryo*, Kregel Publications, 2020.

[197] Strege, *A Snowflake named Hannah*, p. 47 etc.

mother, Marlene Strege. In other words, these children were not the natural children of John and Marlene and the whole book is a testimony to how they went through the options, arriving at the vocation to both adopt and promote the adoption of frozen embryos, inspiring many others to give a home to a child suspended between living and dying. Thus, demonstrating once again, as indeed does every child conceived, that what begins is a "who" and is entitled to the completion of human development and the opportunities that bring out his or her talents and enable both fulfilment and fellowship. John, Marlene and Hannah, and many others, contributed to the growing awareness that their witness to the reality of a child, frozen, thawed, and implanted, means that there are no spare human embryos for experimentation; and, as John, holding two children, Luke and Mark said:

> "'I would like to ask every member of this committee, especially the members that aren't here a question, and that question is: Which one of my children would you kill?" John asked. "Which one would you choose to take? Would you want to take Luke, the giggler, who

we call Turbo, or do you want to take the big guy, Tank? Which one would you take?"[198].

There are no abstract human beings, no potential people; there is each of us, from conception, at the beginning of being unfolded by the dynamic of human development that characterizes human life's manifestation of "who" is there from the beginning.

In view of the history of Israel, a country can have a vocation, that 'all who are on the earth will know that thou art the Lord, the God of the ages' (Sirach 36: 17; but also 1-17). Therefore, just as St. John Paul II saw the role of Poland as a witness to the nations during the appalling occupations of the Second World War and its aftermath[199] so, taking the theme of the American experience as a benefit to all of us, we can see that President Bush expressed one of the wider concerns that witnesses to our common humanity. At the same time, this debate echoed the

---

[198] Strege, *A Snowflake named Hannah*, p. 89.

[199] Cf. there are various accounts of this vocation e.g. George Weigel, *The End and the Beginning: Pope John Paul II – The Victory of Freedom, the Last Years, the Legacy*, Doubleday, 2010.

Nuremburg trial: 'The only difference is that we have substituted human embryos as the group of devalued, commodified human beings who are to be sacrificed on the altar of scientific progress'[200]. Whereas Germany's modern *Embryo Protection Act* of 1990[201] and Italy's *Medically Assisted Reproduction Law*, (Law 40/2004), both 'provide model legislation that ban cryopreservation and

---

[200] Strege, *A Snowflake named Hannah*, p. 94: Dr. Hook, speaking as a private citizen. Cf/ also the various references to the pharmaceutical industry in Bill Bryson's, *The Body: A Guide for Occupants*, Doubleday, 2019; but, also, see the trade-off between the knowledge gleaned from abusive practices and the freedom from accountability of a prominent perpetrator of them, on pp. 228-229: "Unit 731". Indeed, the book contains numerous instances of barbaric medical practices as well as, naturally, the practice and development of good medicine.

[201] Elizabeth Rex's, "An End Word: A New Beginning"; she says: 'in compliance with the Nuremburg Code, passes the world's first "Embryo Protection Act". Under penalty of fines and/or imprisonment, human embryos may not be used for scientific experimentation, harmed or killed' on pp. 597-598, of *Conception: An Icon of the Beginning*: https://enroutebooksandmedia.com/conception/.

Chapter Nine: A New Beginning

all non-therapeutic experimentation human embryos. *An Amended Nuremberg Code must do likewise*'[202].

More generally, then, why does an "industry" have the right to bring about multiple human children and, at the same time, freeze, discard or experiment upon them at will? Where is the outcry? There is indeed a journalistic need to investigate the truth of what constitutes the foundation of a biotechnological company, what its source of funding is and what are the scope of its actual procedures and practices?

Thus President Bush said, with respect to the debate concerning whether government money should be spent on embryo stem cell research which, in effect, ended the life of the particular human embryo from which it was extracted; he said, speaking in the presence of children who were once frozen human embryos:

---

[202] Courtesy of Dr. Elizabeth B. Rex, "Abstract and Power Point Presentation: "International Laws that Promote the Ethical Healthcare and Legal Protection of Human Embyros" etc., prepared for the annual summer conference of the *Center for Bioethics and Human Dignity* (CBHD), Trinity International University in Deerfield, Illinois, June 25, 2022, slides 28-31 on Germany's legislation and Italy's on slides 32-36.

"These boys and girls are not spare parts"[203].

And, on an earlier occasion, President Bush said: 'Extracting the stem cell destroys the embryo, and thus destroys its potential for life'[204].

Let me add, however, that the embryo is already living and, if it were not, the embryonic child would be useless in view of any possibility of extracting embryo stem cells as, it seems, these are taken from a living human embryo.

Thus the embryo is already a living human being and would not bear the brunt of these brutal scientific exploitations if it was not already human. In other words, a

---

[203] Strege, *A Snowflake named Hannah*, p. 22, citing "President Discusses Stem Cell Research Policy", The White House, July 19th, 2006: https://georgewbush-whitehouse.archives.gov/news/releases/2006/07/20060719-3.html.

[204] Strege, *A Snowflake named Hannah*, p. 102 citing "President Discusses Stem Cell Research", The White House, August 9th, 2001: https://georgewbush-whitehouse.archives.gov/news/releases/2001/08/20010809-2.html.

human being's existence, full of the promise of life, is used as a means to an end other than his or her own good; and, if this is possible for the least of human beings, is it not possible that we are all vulnerable to that logic: that our lives are expendable if another, or others, deem there to be an end that our death is the means?

> 'At its core, this issue forces us to confront fundamental questions about the beginning of life and the ends of science'[205].

Certainly, we need to answer truthfully the question of the 'beginning of life', the question that each of one of us has answered in the very existence we live: that there was a point of beginning from which our identity unfolded to today. So, there is no question of the alienation of science or scientists; rather, there is the guidance of research by the obligation upon us all *to do good and avoid harm*. Thus, in the end, we are searching for the truth that will obtain peace between us, as Pope Francis says: "Peace results from an enduring commitment to mutual dialogue,

---

[205] Strege, *A Snowflake named Hannah*, p. 103, citing "President Discusses Stem Cell Research" etc.

a patient search for the truth and the willingness to place the authentic good of the community before personal advantage'[206]. Thus the 'authentic good of the [whole human] community depends upon recognizing our common humanity's original moment of beginning and subsequent development.

An ITEST Webinar Entitled "A Post-Roe World"[207] (I).

We now turn to other new sources, beginning with the webinar entitled "A Post-Roe World", entailing two presentations, one by Dr. Pat Castle and another by Kiki Latimer; they presentations complemented each other beautifully although, as both of them are practical helpers,

---

[206] ADDRESS OF HIS HOLINESS POPE FRANCIS TO THE PARTICIPANTS IN THE MEETING PROMOTED BY THE INTERNATIONAL CATHOLIC LEGISLATORS NETWORK, Clementine Hall, Thursday, 25 August 2022: https://www.vatican.va/content/francesco/en/speeches/2022/august/documents/20220825-cath-legislators-network.html.

[207] Go to: "A Post-Roe World with Dr. Pat Castle and Kiki Latimer": https://www.youtube.com/watch?v=ms4NEi56ZPY.

Chapter Nine: A New Beginning

they naturally overlap each other too. Together, however, they wonderfully synthesize the work of both explaining the reality of abortion and, at the same time, help us to understand what help is available and, in the case of counselling, how subtle and sensitive that help needs to be.

Dr. Pat Castle: '*78% of post-abortion mothers said if they had encountered ONE supportive person or encouraging message, they would have chosen life.*' And, therefore, he founded "Life Runners": they wear '"REMEMBER The Unborn" jerseys as a public witness in over 3,300 cities'[208]. In other words, given the dearth of interest in the media advertising the help that exists, it is a tremendously practical exercise to signal that that help exists; and that help, to take the American situation as illustrative of what is offered, can be very extensive: '79% of post-abortion

---

[208] ITEST "A Post-Roe World" Webinar publicity, now with the WCAT TV presentation on it: https://faithscience.org/post-roe/.

mothers didn't know about free help. Lead mothers to assistance at one of the 3,000+ pregnancy help centers'209.

Kiki Latimer focused, then, on the pastoral care of a woman as she came for help and, having accumulated considerable experience, her presentation and the ensuing discussion well illustrate what she said she was going to talk about:

'Looking at the primary reasons women choose abortion helps us to find avenues for prevention of abortion as well as healing after abortion'

And:

'Breaking down one or more of these barriers during a crisis pregnancy can help a woman choose life for her baby and herself. After an abortion, a recognition of the one or more of these barriers that led to the abortion can help in the transformation to healing and reconciliation.

---

209 "Facts of Life": https://www.liferunners.org/facts/; and mothers are led to this link: "Help and Healing": https://www.liferunners.org/help/.

Chapter Nine: A New Beginning

Finding ways to reach women (and the connected men) in these circumstances requires listening skills, compassion, and a willingness to understand how they see the situation'[210].

In one of her touching examples, a young woman, pregnant and seemingly convinced of the need for an abortion, came into her office. After Kiki had listened to several reasons why the young woman needed an abortion, she was given a cuddly toy which transformed a rather abstract discussion about being pregnant into a very concrete moment in which the young woman recognized that she was carrying a child, her child, and decided to keep her baby[211]. In a different, but parallel moment of recognition, a woman who was processing data from an *in-vitro* fertilization clinic realized, when she became pregnant, she was no longer processing "data" but babies

---

[210] ITEST "A Post-Roe World" Webinar publicity, now with the WCAT TV presentation on it: https://faithscience.org/post-roe/.

[211] Cf. WCAT TV presentation: https://youtu.be/ms4NEi56ZPY.

and she left and began a ministry for those who died in the IVF process[212]. In other words, one of the simple but profound challenges is helping people to see that if a woman is pregnant or a child comes to exist, that this is the start of a relationship to another human being, whether son or daughter, and that there is an immense range of help available to those who need it.

An ITEST Webinar Entitled *"Bioethics and Law: Understanding* **the** *Nexus: Truth and Meaning in Constitutional Jurisprudence"*[213] **(II)**

We now move on to another webinar, "Bioethics and Law", with Fr. Thomas David, which gives an overview of the American Constitution and various questions

---

[212] Cf. Laura Elm's contribution, "Foreword to Chapter Three", to *Mary and Bioethics: An Exploration*, pp. 105-110: https://enroutebooksandmedia.com/maryandbioethics/.

[213] WCAT TV presents . . . ITEST Webinar with Fr. Thomas Davis on Bioethics and Law: Understanding the Nexus: Truth and Meaning in Constitutional Jurisprudence: Bioethics, Babies & Bromides": https://youtu.be/928snYsmT3k.

Chapter Nine: A New Beginning

that have arisen from it. Fr. Thomas David 'is the Founder and Director of the Liberty Institute for Faith & Ethics (LIFE) a multidisciplinary center dedicated to scholarship and study of cultural, legal, and bioethical issues. Its Bioethics Library and Religious Liberty Observatory offer extensive collections of scholarly articles, primary source documents, and life affirming medical directives related to bioethics and religious liberty. LIFE's website and it Bioethics Library may be accessed at liberty4life.org.'

In Fr. Thomas' excellent and complementary presentation on the background to the American Constitution and the foundational judgements of the Supreme Court which, over time, have clarified what is and is not entailed in the Constitution and, moreover, what are the respective powers of the Government and the States of America, allows the emergence of wider questions that pertain to bioethics and law more generally. Nevertheless, to complete a brief account of the American situation, there follow a few general comments and a note about their wider implications.

In general, a principal response of the Supreme Court's judgement that there is no Constitutional origin to a woman who wants to abort her child is that at the time of the Constitution, the States had legislated on behalf of the unborn, albeit with variations. In the first one, enumerated chronologically by date, Missouri, in 1825, said that, with respect to the abortion of a child, that a person is guilty of murder if it is proved that it was done, by whatever means, 'with an intention to harm him or her thereby to murder,'[214] the unborn child. In other words, as Fr. Thomas notes in the course of his presentation, there are already historical grounds for considering, as some argue, that the child was already regarded as a "person" at the time of the American Constitution which, if this can be demonstrated clearly, then the child is protected by the American Constitution; and, as this statute uses the language of 'with an intention to harm him or her', it can hardly be misunderstood to mean that the child is a person in so far as it is ordinarily understood that a person is 'him or her'. Indeed, more generally, the word 'child'

---

214 P. 79 of "Appendices A": https://www.supremecourt.gov/opinions/21pdf/19-1392_6j37.pdf; and, in general, these references to 'him or her', the 'child' (which is so much more common) and the exemption 'to preserve the life of such mother', are drawn from pages 79 – 108.

Chapter Nine: A New Beginning

occurs and recurs with the natural understanding of a child of the parents and a member of their human family and, as understood, a person in every sense except that of admitting them to American citizenship until they are born. Moreover, as with the wording of the Missouri statement and indeed many others, 'person' is reserved for the one who may be 'guilty of murder', while otherwise 'mother' and 'child' are referred to in terms of their relationship to one another; and, therefore, there is an equality of understanding that 'mother' and 'child' are no less persons – they are simply referred to in terms of their natural relationship.

But, if the intention is to admit a child to American citizenship on birth, then it follows that the child is understood to be a person and not an animal, plant or less than human being; as it says in the 14th Amendment: 'All persons born or naturalized in the United States, and subject to the jurisdiction thereof, are citizens of the United States and of the State wherein they reside.'

Notably, then, by 1919, the definition of the life of the child is altogether more comprehensive; and, therefore, in the statute for New Mexico it says: 'Sec. 3. "For the purpose of the act, the term "pregnancy" is defined as that condition of a woman from the date of conception to the birth of her child"[215]. I would also comment that as any number of these statutes allow the exception against prosecution in view of the clause, 'unless the same [referring to a deliberate abortion] shall have been necessary to preserve the life of such mother', a corollary of this would be that, as medicine develops, there is an obligation to act in such a way that the life of the child is protected along with that of the mother – if, that is, the life-threatening condition of the mother can be helped without endangering the life of the child. More generally, the wording is so similar between one state, whether putative or not, that a common understanding is entailed: that of protecting the mother and child unless an intervention is necessary for the life of the mother which, as we know in our own times, is less and less actually necessary.

In answer to the specific question: Is there a 'legal definition of the human person which could protect the identity of the human race from animal-human hybrid, non-therapeutic experiments, being frozen or discarded?' Fr. Thomas answered that there is a genetic definition of the human person. Taking account, then, of the work which establishes human identity on the basis of human

---

215 "Appendices A", p. 107.

embryology[216], it is clear that, acknowledging the variations which are characteristic of human beings, such as an additional chromosome, we have the basis of what is a universal definition of human identity – unless obscured by animal-human experiments. All in all, then, personhood is understood on the basis of genetic membership of the human race; rather then, a condition or characteristic that can be established independently of it. In other words, a human being is a human person.

The ongoing challenge: Is there a right to life in the American Constitution?

Currently, then, taking up what seems to be a distinct possibility in the understanding of the American Constitution's 14th Amendment, 'Catholic pro-life advocates formally asked the U.S. Supreme Court this month to recognize the personhood of unborn babies under the U.S.

---

[216] Cf. "Chapter 5: Part II", written by two specialists: Professor Justo Aznar (a former member of the Pontifical Academy for Life) and Julio Tudela for Etheredge, *Conception: An Icon of the Beginning*: https://enroutebooksandmedia.com/conception/.

Constitution. The push to recognize fetal personhood has the potential to produce a nationwide abortion ban if the Court affirms that the unborn are persons with guaranteed rights'[217]. However, 'The Supreme Court has declined to hear a case from Rhode Island about whether unborn babies have a right to life'[218]. The point remains that a child becomes a citizen on birth; and, by implication, as citizenship is granted to a person, the preborn child is therefore understood to be a person.

'Professor John Finnis at the University of Notre Dame Law School and Professor Robert P. George at

---

[217] Ashley Sadler, September 10th, 2022, "'Now is the time': Catholic pro-life group petitions US Supreme Court to recognize fetal personhood": https://www.lifesitenews.com/news/now-is-the-time-catholic-pro-life-group-petitions-us-supreme-court-to-recognize-fetal-personhood/?utm_source=digest-prolife-2022-09-13&utm_medium=email.

[218] Steven Ertelt, October 11th, 2022, "Supreme Court Won't Hear Case Declaring Unborn Babies Have a Right to Life": https://www.lifenews.com/2022/10/11/supreme-court-wont-hear-case-declaring-unborn-babies-have-a-right-to-life/.

Chapter Nine: A New Beginning

Princeton University filed an amicus brief in the Dobbs case presenting a detailed history of American law that, according to their research, recognized unborn babies as persons under the Fourteenth Amendment up until *Roe v. Wade* in 1973. In their brief, Finnis and George refuted the idea that the U.S. Constitution is "silent" on the matter of whether an unborn baby is a legal "person" under the Fourteenth Amendment'[219].

Thus the investigation of the 14th Amendment continues.

A possible comparison between the legislative variations of American States and a precept of the European Court

Without discussing the details of this case, there is nevertheless a relevant legal doctrine identified by the European Court, that may be advantageous to the progress of legislation across the world that defends the integrity of human life from the first instant of fertilization:

---

[219] Ertelt, October 11th, 2022, "Supreme Court Won't Hear Case Declaring Unborn Babies Have a Right to Life".

'As justification, ... [the European Court] cited *the margin of appreciation*, a legal doctrine which allows particular countries a degree of freedom in the practical implementation of human rights'[220].

At the same time, In ... [Judge Vincent A. De Gaetano's] view, human procreation had been reduced from an act between a man and a woman to a medical or laboratory technique. Despite medical progress, recognition of the value and dignity of each and every human being may well require the introduction of a ban on some activities in order to bear witness to the sacrosanct value and inherent dignity that concerns each and every human life. Such a ban would not be a

---

[220] Appendix: Case Illustrations, italics added to the quotation on p. 123, but see also p. 130 of Alichniewicz and Monika Michalowska "Medicine of the Beginning of Life." Etc. For the source on p. 130 go to: European Court of Human Rights, Case of Parrillov. Italy, Judgment, August27, 2015, http://hudoc.echr.coe.int/eng?i=001-157263 (accessed October 15, 2019) – as per the author's document.

denial of fundamental human rights but rather their recognition'[221].

What is more, given the following wording of another judgement of the European Court, there is grounds for the beginning of new developments in terms of defining a human embryo:

The European Court 'argued in its justification that while the notion of property could not be limited merely to material objects, human embryos could equally not be reduced to the level of an object'[222].

While, then, there can be a problem of contradiction if a law grants protection to the human embryo, which

---

[221] *Ibid*, pp. 123-124: European Court of Human Rights, Case of S.H. and others v. Austria, Judgement, November 3, 2011, http://hudoc.echr.coe.int/eng?i=001-107325 (accessed Oc-tober 15, 2019) – as per the author's document.

[222] *Ibid*, p. 131: European Court of Human Rights, Case of Parrillov. Italy, Judgment, August 27, 2015, http://hudoc.echr.coe.int/eng?i=001-157263 (accessed October 15, 2019) – as per the original author's document.

cannot 'be reduced to the level of an object' but makes exceptions, exceptions, as we have discussed, may become redundant in due course owing to the improvement of interventions to the benefit of both mother and child; and, what is more, beneficial care could be made available across the world: a kind of "beneficent" clause that would oblige, in law as well as in charity, that countries with advanced medical care help where it is crucially necessary in countries which are less developed.

Coercion: Personal and Social

In addition to all the difficulties of addressing the help that people need, with respect to considering the possibility of an abortion, is the reality of coercion; but coercion is both personal and social.

Personal Coercion

'A recent BBC poll found that 15% of women aged 18-44 reported they had been pressured into having an abortion they did not want - that is 15% of ALL

pregnant women - not just women who have had abortions'[223].

In a vivid account of what happens, we read 'After taking, on … [the boyfriend's] instructions, pills bought online (he was opposed to her going to a doctor), Laura delivered a dead baby, with much loss of blood. She is quoted as saying "I wanted to die. Honestly, I just felt like the whole world had just ended in front of my eyes"[224].

In other words, quite apart from misinformation about the reality of human conception, that it is not a "who" who comes to exist but some kind of "blob of cells", the multitude of problems that pregnancy may present, there is the

---

[223] "15% of UK Women Coerced into Abortions": http://www.cmq.org.uk/CMQ/2022/Aug/coercion_of_abortion.html.

[224] Helen Watt, 11th October, 2022, "Tragic stories of abuse should not lead the UK to decriminalise abortion": https://mercatornet.com/tragic-stories-of-abuse-should-not-lead-the-uk-to-decriminalise-abortion/81183/.

"pressure" applied to a person who is already struggling, in one way or another, to accept the life within her.

Another woman, Hayley, speaks very clearly about both what happened to her, at 16, and the post-abortion syndrome that went on for 21 out of 24 years, "lost and broken, suicide felt like a really good choice for me", entailing a spiralling down into drugs, drink, eating disorders and promiscuity, until she met Christ and was forgiven – and now she wants to give others the hope of His healing help too – as well as calling on the government to listen to the reality of women's experience[225].

Social Coercion

In what sense, then, can this be called a "pro-choice" situation when the culture, the technology, the law and the media, are so clearly weighted against both mother and child: against one of the basic, bonding relationships at the root of human existence? Indeed, as Bishop John Keenan

---

[225] Cf. "Hayley: Abortion and Coercion: Speaking in Parliament Square", September 6th, 2022: https://youtu.be/X01bZlNNF0c; and she says that Rachel's Vineyard (https://www.rachelsvineyard.org.uk/) and "Her Voice".

suggests, there is a very real danger that free speech is under threat as it becomes socially unacceptable to protest, peacefully, to pray and counsel outside an abortion facility or to otherwise raise the contrary point of view in the pursuit of truth[226].

Just because the Warnock Report could not see the moral implication of the continuous process of development from conception onwards, implying that this truth be the foundation of a categorical ban on the exploitation of human beings, does not mean that the arbitrary claim to destroy a human being up to a certain period of time can remain unrevised – for that human being is now irrevocably dead. On the contrary, that very contradiction between the truth of an uninterrupted unfolding from the first instant of fertilization, of itself, embodies the good to be done: the recognition of a human life begun. And, therefore, restoring the relationship between truth and doing good[227], entailed in recognizing the rights integral

---

[226] Cf. "Uplift: Bishop John Keenan", September 14th, 2022: https://www.youtube.com/watch?v=M8NCAYIFoCc.

[227] Cf. Francis Etheredge, "On Regulating IVF" in "Ethics and Medics", Volume 41, Issue 7, July 2016. Similarly, there is hope of the repeal of the UK 1967 Abortion Act, along with the

to human relationships from the beginning, simultaneously restores the teaching power of the law: teaching the equal inviolability of each and every one of us.

So, just as America is seeing a revision of its laws, in accordance with sound reason and scientific evidence, so the hope is that other countries will follow, albeit according to their own legislative paths and, who knows, maybe the unthinkable will be thinkable: a renewal of abiding, bonding, international agreements on the beginning of human life and the full unfolding of the person present from conception – or from whatever point of beginning he or she came to exist, energetically and dynamically propelling the irrevocable manifestation of the person begun.

---

repeal of the Human Fertilization and Embryology Act, 2008. Moreover, Australia has abortion up to birth in all states; read "The abortion conversation we need to start having", by Joanne Howe, September 2nd, 2022: https://www.catholicweekly.com.au/the-abortion-conversation-we-need-to-start-having/; but, also, read between the lines and imagine what an abortion up to birth really means: What concrete actions does it entail? Why aren't these children offered for adoption?

Chapter Nine: A New Beginning

An Ectopic Pregnancy: Fear, Truth and Opportunity

Of the many difficulties that women face in becoming pregnant, one is particularly sensitive: the child coming to exist and starting to develop – but not within the womb, and commonly within the fallopian tube, the tube that carried the ready to be fertilized egg from the mother's ovary. One of the fears is that legislative protection of the child conceived ectopically, is that it is unrealistic:

'Ohio's bill instructing doctors to "reimplant" an ectopic pregnancy is part of a worrisome trend of lawmakers inventing unproven therapies related to reproductive health, say Daniel Grossman and Yanett Anaya'[228].

---

[228] "The myth of ectopic pregnancy transplantation", December, 2019: https://blogs.bmj.com/bmj/2019/12/17/the-myth-of-ectopic-pregnancy-transplantation/; cf. also Jessica Glenza, 29th November, 2019, "Ohio bill orders doctors to 'reimplant ectopic pregnancy' or face 'abortion murder' charges": https://www.theguardian.com/us-news/2019/nov/29/ohio-extreme-abortion-bill-reimplant-ectopic-pregnancy?CMP=Share_iOSApp_Other.

'In addition to ordering doctors to do the impossible or face criminal charges, House Bill 413 bans abortion outright and defines a fertilized egg as an "unborn child"'[229].

However, before discussing the current situation with respect to ectopic pregnancies, it is necessary to address the definition 'that a fertilized egg [is] … an "unborn child"'. As we will see in the course of this book, a 'fertilized egg' is both child and what that child needs to implant in the womb and benefit from it; just, then, as a parachutist jumps out of an aeroplane, would we not distinguish between the man or woman and the parachute – such that when they land, we would separate the chute from the person. Likewise, with underwater divers, when they surface, we have no difficulty distinguishing man or woman from oxygen tanks and diving suit. To take a more organic example, the courgette seed has a very visible seed case which, while it was integral to the development of the seed, is discarded as the embryonic plant grows.

---

[229] Glenza, 29th November, 2019, "Ohio bill orders doctors to 'reimplant ectopic pregnancy' or face 'abortion murder' charges".

Similarly, therefore, when it comes to the integral beginning of the human person, embryonic development demonstrates a differentiation between the child and what is necessary for intrauterine life. When a child is born, there is no confusion about what is the umbilical cord, placenta and embryonic sac, enabling the child to survive in a liquid environment – and the child. What, in the end, is the point of the definition 'that a fertilized egg [is] ... an "unborn child"' if not that child and everything that pertains to the intrauterine life of that child is entailed in the 'fertilized egg'; and, indeed, if this were not the case, why would anybody get pregnant, fertilize a human egg, implant the child or freeze it? In other words, it is incoherent to deny or quibble over the relationship between fertilization of a human egg and the presence of a child.

With respect to ectopic pregnancies, then, there would be no dramatic concern if there were not mother and child intimately involved in the crisis pregnancy. Thus, considering the urgency of the situation, it is necessary to address a number of salient points. One of the first successful, surgical helps, to both mother and child, was in 1917:

'A doctor named C.J. Wallace thought it was possible and proved it in 1917. Wallace performed the first successful attempt on record to transfer an embryo in an ectopic pregnancy to the uterus, saving both mother and child'[230].

So, it must be asked, why has there been so little interest in saving the life of both mother and child? Why, also, are there so many pregnancies of this kind? In America, for example,

'About one in 50 pregnancies in the U.S. today are ectopic, meaning the embryo implants in a location other than the uterus'[231].

---

[230] Margaret Peppiatt, July 22nd, 2022, "Treating ectopic pregnancies is not abortion. But researchers are still looking for a way to save both mother and child": https://www.americamagazine.org/politics-society/2022/07/22/embryo-transfer-research-ectopic-pregnancies-243400; in the same article there was a second, successful transfer, in 1990, by Dr. Landrum B. Shettles, of a 40 day old child into the womb of the mother.

[231] Peppiat, "Treating ectopic pregnancies" etc.

Chapter Nine: A New Beginning

Against this unlikely background, albeit involving two actual, successful transfers, there is a new enthusiasm for helping both mother and child:

'Based on his findings so far, [Dr.] Sammut [who is a Professor at the Franciscan University of Steubenville] said he believes that "embryo transfers have the potential to become a standard treatment of ectopic pregnancies in the future."'

On the one hand, given the plight of both mother and child, and the claims of million-dollar industries to be of help to women who would like the gift of child: Why has there been so little concrete interest in the problems entailed in ectopic pregnancies – of a child conceived outside the womb? Is it really true that the lack of monetary rewards accounts for the dearth of interest from these "powerful" biotechnological companies[232] or research institutes? If so, how profoundly true it is that our culture has "lost" a sense of relationship: that the bonds between us are "cashed" in for money?

---

[232] Peppiat, "Treating ectopic pregnancies" etc.

'Among the patients who become pregnant after assisted conception, around 4% of the pregnancies will be ectopic. The embryos migrate to the ostial ends of the tubes after transfer, or they may inadvertently be placed there when they are transferred. Heterotopic pregnancy (a multiple pregnancy with one embryo in uterus and one in the tube) is extremely rare with natural conception, but the rate maybe as high as 1% in assisted conception'[233].

On the other hand, what if the use of an intrauterine device, or IUD, is connected to a risk of ectopic pregnancy: 'one 2004 study (cited by ACOG[234]) found that over half (53%) of pregnancies that occurred in women with an IUD were ectopic'[235]. In other words,

---

[233] Cf. Violetta Anastasiadou and Joep Geraedts, et al, p. 606 of "The interface between assisted reproductive technologies and genetics" etc.

[234] ACOG is the acronym for the American College of Obstetricians and Gynecologists.

[235] Grace Emily Stark, July 21st, 2022, "Your treatment options for an ectopic pregnancy, and why the overturn of *Roe* hasn't affected them": https://naturalwomanhood.org/

what if the lack of interest in ectopic pregnancies is because of its connection to contraceptive practices and the industry that profits from them?

It is debateable, however, that the way forward is to address the plight of the child conceived ectopically through legislation; however, by bringing the whole dramatic need, of both mother and child, into the limelight, it is certainly possible that, as we have seen, it may both raise challenging questions and stimulate positive developments – *for both mother and child!* After all if, as it is claimed, medicine is for the good of all – then let us hope that the good of all prevails!

Pope Francis on Justice, the Bond of Fraternity and Peace

Pope Francis, in his *Address to Catholic Legislators*, develops three points that are applicable to our discussion, indeed applicable more widely than just to Catholic

---

how-do-you-treat-ectopic-pregnancy/?mc_cid=3c0395956c&mc_eid=b1f4a58a67.

Legislators, in that these three ingredients encourage a more humane understanding of what is involved in framing laws. Pope Francis says: '*justice*, classically defined as the will to give to each person what is his or her due...'.

And what is due to a person who is conceived if not that the natural process of human development, once irrevocably begun, continues the unfolding of the person present from the beginning?

> But where is justice without fraternity? 'In fact, a just society cannot exist without the bond of fraternity, that is, without a sense of shared responsibility and concern for the integral development and well-being of each member of our human family'.

So, what will make possible that 'bond of fraternity' if we cannot recognize each other's humanity; and, therefore, abuse the gift of life which is equally given to each of us, establishing our equality before the law?

What, then, unfolds from the truth of human conception and the bond of human fraternity, if not the peace that the human community so desperately needs to flourish in the times in which we live:

Chapter Nine: A New Beginning

'Peace results from an enduring commitment to mutual dialogue, a patient search for the truth and the willingness to place the authentic good of the community before personal advantage. In such an effort, your work as lawmakers and political leaders is more important than ever. For true peace can be achieved only when we strive, through far-sighted political processes and legislation, to build a social order founded upon universal fraternity and justice for all'[236].

Practically, then, this leads both to laws that express our common humanity but also to those practices, like adoption[237], which make concrete the love of another,

---

[236] Pope Francis, Address to Catholic Legislators, Clementine Hall, Thursday, 25 August 2022: https://www.vatican.va/content/francesco/en/speeches/2022/august/documents/20220825-cath-legislators-network.html.

[237] Pope Francis Encourages Adoption and Warns of "Demographic Winter": http://www.cmq.org.uk/CMQ/2022/Aug/pope_francis_encourages_adoption.html; and, while less well known for this, President Obama 'listed a number of proposals to reduce the number of abortions, including "Let's make adoption more available!"': Charlie Camosy, 7th October,

whether that child is a frozen embryo or otherwise in need of a home.

2022, *The Pillar*, "Supporting adoption in a post-Dobbs America": https://www.pillarcatholic.com/supporting-adoption-in-a-post-dobbs-america/.

Chapter Ten

A Conclusion in Three Parts:

Part I: Lest We Forget Mother, Child and Father; and

Part II: The Wider Implications for a Post-Roe World; and

Part III: The Disruptive Power of The Word of God

Part I:
Lest We Forget Mother, Child and Father

This is not an abstract discussion although, at times, it takes us into the most difficult philosophical terminology that there is, the most amazing and detailed analyses of embryology and the most intense and controversial social disagreements in modern times; it is a discussion about specific people, whether born, under the heart of the mother or frozen: it is about who has received the gift of human life is simply equal to anyone else who has received the gift of life. Thus no one is excluded from the world discussion of what we, as human beings, are bringing about in the present ethical climate of the human race. As we emerge, then, from our national identities and increasingly recognize that abstract truths about human personhood, that to be a human person is to be a human being-in-relation, need "returning" as it were to the concrete reality from which they came – we will appreciate more and more that mother and father, parent and child, brother and sister, aunt and uncle communicate the profoundly interpersonal structure of human identity.

We have discussed various ways to understand human being, drawing on analogies or comparisons so that what is unfamiliar becomes intelligible and familiar; but, in the end, as we make actual progress in understanding the true reality of human conception, so we can see that our comparisons have strengths and weaknesses and that we are called, more and more, to communicate the truth that we have discovered: the truth to be "embodied" in a universal declaration of human rights. Thus, there is a mother who needs help[238]. There is a child who needs help. There is a father who needs help to understand his fatherhood[239]; indeed, fatherhood is resoundingly understated and

---

[238] Myers, "Law Professor Reflects on Landmark Case": https://ndsmcobserver.com/2013/01/law-professor-reflects-on-landmark-case/.

[239] Cf. Francis Etheredge, *The Prayerful Kiss*, particularly: "Indelible". But in a joint, bioethical investigation, the authors say: 'A possible explanation [for the lack of data on men with respect to procreation] can be provided by the fact that, as some authors indicate, there has been a general social tendency to identify reproduction issues as a 'women's matter' (Chapter 3, p. 107 of Alichniewicz and Monika Michalowska's, "Medicine of the Beginning of Life." Etc).

almost totally absent from the whole discussion. Why is this?[240]

And there is a world of people with vested interests who need help to appreciate that all human beings are a gift, equally given into the care of each of us; and, therefore, that whatever the good intended, there is an obligation on everyone to understand that there is an objective good for each and every one of us, without exception, otherwise there is an actual inequality between us. And if there is an actual inequality between us, as regards who is a human being, then there is the increasing possibility of an exponential increase in the exploitation of all of us; for human rights, in the end, are the rights of relationship: the rights of relationship which come into existence when human beings are conceived – conceived in relationship to the whole human race and to God.

Just, then, as a plant exists in an ecological context and what is done to it, for better or worse as regards the health of the individual plant and its place in the ethico-eco-

---

[240] Cf. the rise of articles in The Catholic Weekly, Sydney, on fatherhood e.g. Daniel Ang, September 16th, 2022, "The love of a father": https://www.catholicweekly.com.au/the-love-of-a-father-2/.

system, so what is done to individual human beings impacts, for better or worse, on the "ethical" whole of the integrity of the human community. But, 'Conscious' as we are of our 'limitations', we ask 'for constant prayer and intercession for' our 'efforts to translate the sacred texts "in the same Spirit by whom they were written"'[241]; and, therefore, we need to ask for the same help to read both the book of nature and all that helps us to recognize what is universally good, true and applicable to all. Thus, in a spirit of universal fraternity, any progress in the Church's teaching on conception can only complement any development in the natural understanding of it; and, in that same spirit, one source can complement another while being very different from it. Thus, in the end, there is the progress of the hope that international agreements and laws will not only help the plight of people in the present but will lead to the protection of our common humanity; indeed, is it time that the "human race" becomes the

---

[241] Pope Francis, *Scripturae Sacrae Affectus*, commemorating the life and work of St. Jerome, citing *Praefatio in Pentateuchum*: PL 28, 184': http://w2.vatican.va/content/francesco/en/apost_letters/documents/papa-francesco-lettera-ap_20200930_scripturae-sacrae-affectus.html#_ftn19.

"subject" of human rights and international agreements and, ultimately, international bioethical law?[242]

Putting it at its simplest, children are genetically members of the human race, irrespective of variations and, therefore, the biological foundation of human personhood is from conception onwards, psychological development being expressed as intrauterine development naturally manifests it[243]. At the same time, the parents look

---

[242] Elective abortion is by definition discriminatory against the life of the child aborted; and, therefore, if the law progresses through the rejection of what is discriminatory, then it is right to reject elective abortion; cf. also "Arguments from International Human Rights Law in the Supreme Court's Dobbs v. Jackson Women's Health Organization Abortion Case" by Alexis I. Fragosa, Esq: https://c-fam.org/definitions/arguments-from-international-human-rights-law-in-the-supreme-courts-dobbs-v-jackson-womens-health-organization-abortion-case/.

[243] Francis Etheredge, Chapter 6, "A Three Part Exploration of the Human Being as a Psychologically Inscribed Biology: Part I: An Indivisibly Psychosomatic Being from Conception; Part II: A Pilgrimage from Reality to Self-Knowledge and the Challenge of the "Inexpressibility" of the Human Person; Part III: Embodied Relationality and the Integral Nature of a

forward to meeting their child and this is intrinsic to the suffering experienced in infertility, miscarriage and regret after abortions; and, as such, is profoundly significant in the recognition of the "relationship" which is inseparable to the parent-child co-existence.

Note the contrast between two losses, the first from processes that were meant to ensure a healthy child and the second from a father of a child who died of cancer just after his second birthday; but, who knows, maybe the first group share the reaction of the father below but it is hidden amidst other reactions.

'The parents may claim that, as a result of the fault, they have been deprived of the opportunity to eliminate or terminate the pregnancy and they are burdened with a sick or handicapped child'[244].

Psychologically Inscribed Biology" of *The Human Person: A Bioethical Word*: https://enroutebooksandmedia.com/bioethical-word/.

[244] Violetta Anastasiadou and Joep Geraedts, *et al*, p. 611 of "The interface between assisted reproductive technologies and genetics" etc.

In contrast to the silence and loss of so many children is the pain, so beautifully expressed, by a father of his third child who died of brain cancer:

'The truth is, despite the death of my son [He had just turned two[245]], I still love people. And I genuinely believe, whether it's true or not, that if people felt a fraction of what my family felt and still feels, they would know what this life and this world are really about'[246].

May the Lord make it possible for the pain of loss to be fruitful!

---

[245] Rob Delaney, Extract from *A Heart That Works*, published by Coronet, 2022, cited on p. 13 of *The Sunday Times Magazine*, pp. 8-15, 16th of October, 2022.

[246] Rob Delaney, Extract from *A Heart That Works*.

Part II:
The Wider Implications for a Post-Roe World

There will no doubt be ongoing controversies about the natural right of members of the human race to be conceived free of animal-human hybrid experiments[247] and the right to integral, completing human development: so that once a child is conceived, he or she can be protected from exploitation, rescued from the frustration of being

---

[247] In a brief correspondence with Roberto Andorno, he said: 'But I don't think that human rights can really help to address the kind of issues that you mention (for instance, a right not to be conceived as a genetically modified human being) because human beings who do not exist yet -who have not even been conceived yet- cannot have any human rights... We need some other legal-conceptual tools that relate, not so much to existing individuals, but to the integrity of future generations, and the "human condition" in general' https://www.academia.edu/Messages?atid=18910888). Thus this raises the question of not just the future of the human race – but the human race as the real and substantial "subject" of human rights.

frozen and, finally, protected from the whole field of destructive experimentation, the vested interests of investigators, multi-nationals, harvesting of organs and a whole, almost unimaginable world of exploiting the human being as a "resource" – for this human being is our brother or sister. To recognize the human face of a once frozen embryo, called Hannah Strege, means to be unable to hide behind technological, embryological or historically conditioned definitions of human beings coming to exist, just as it is to recognize that every day, whenever and wherever a child is conceived, there were you and I at the very beginning of our existence.

We stand, then, at a point in human history where it is not so much a question of personal choice determining anything and everything as choosing the truth, as it becomes more fully known concerning human conception, that will take us into a humane future of the human race or the future of the human race will be determined by the most powerful and prevailing vested interests that will determine, on utilitarian grounds, whose future it will be to be a resource for the rest of the human race. If there are documents explaining the ethical relationship to one another, because of our equality as human beings[248] and the tragic events of our times, then how much more necessary is a revisiting of these foundational expressions in the light, or flickering light, of the times in which we currently live.

Thus, whatever the difficulties, we need a workable definition of human personhood, capable of being the basis of a stable interpretation of international law. If, then, it is possible to define the existence of a "Juridical Person" in the following terms:

> 'A juridical person is a non-human legal person that is not a single natural person but an organization recognized by law as a fictitious person such as a corporation, government agency, NGO or International (intergovernmental) Organization (such as United Nations)'[249] –

---

[248] Cf. Chapter 5: Part II, a masterful review of the biological evidence concerning conception written by Profs. Justo Aznar Lucea and Julio Tudela in Etheredge, *Conception: An Icon of the Beginning.*

[249] "Juridical Person": https://en.wikipedia.org/wiki/Juridical_person.

then it follows that human intelligence, creativity and legal ingenuity can devise, for the sake of our ongoing humanity, a legal definition of human being that could, possibly, be the basis of international law. Incidentally, the expression, 'juridical person' is already in use with respect to the human embryo: '[In] 1986 [Louisiana Health Law, Chapter 3] ... established [the] "legal status" of an IVF human embryo as a "juridical person"[250].

If, then, we have learnt anything from the examination of the American Constitution[251] and the respective cases, by which the Supreme Court has come to a stable interpretation of it, we have seen the benefit to a people, and

---

[250] Courtesy of Dr. Elizabeth B. Rex, email, 13/09/2022, "Power Point Presentation: "International Laws that Promote the Ethical Healthcare and Legal Protection of Human Embyros" etc., prepared for the annual summer conference of the *Center for Bioethics and Human Dignity* (CBHD), Trinity International University in Deerfield, Illinois, June 25, 2022, slide 37.

[251] The question of the revision of the American Constitution, raised by Fr. Thomas Davis, is beyond the scope of this discussion: https://youtu.be/928snYsmT3k.

therefore not just to a people, but to all of us, of the right rule of law; however, just as progress can be made on establishing what a constitution does not entail, so further progress can be made about what it does entail. If it is possible that a thorough going investigation of the American Constitution highlights the possibility of human personhood being implicitly understood, in that nothing less than a person could be granted citizenship on birth, then it is possible that the world community can recognize the historical antecedents to a contemporary recognition of the international nature of the human citizenship of the world.

Let us hope, then, that love will be informed by the truth and establish a charitable understanding of what it is to be a human person, a human being-in-relationship, that will be of benefit, not just to us but to the historical development of the integrity of the human race.

Three Loci: To know our Identity; a Move Towards the Further Denial of Being Human; and Change

On the one hand, the move towards people who are conceived on the basis of donor eggs or sperm, and who naturally have a desire to know who the man and woman were, even if the people who have actually brought them into existence and look after them are, one hopes, the more immediate and real parents. The human desire to know our history brings to the fore the discrepancy between people who manipulate the lives of others[252], literally, and the boy or girl conceived and the needs of their human identity. In other words, to be conceived, ordinarily, through a husband and wife who know each other, is far more significant than was realized, precisely because there is a whole human history which comes with the child conceived: parents; grandparents; and extended family. This is one more piece of evidence which, yet again, shows the priority of the human identity of the person conceived over the artificial methods that "so-called" scientists are using.

On the other hand, there are moves to change international declarations by a sleight of hand, as it were, claiming abortion to be a right rather than demonstrating that

---

[252] "Children conceived by gamete donation in France will have the chance to know their biological origin", by Bioethics Observatory | September 5th, 2022: https://bioethicsobservatory.org/2022/09/children-conceived-through-gamete-donations-in-france-could-know-their-biological-origin/44204/.

it is one[253]. In this debate, it is not the right itself which is demonstrated, as that the claim is insistently made that women need access to abortion: to end a human life within the womb. In specific cases, as we have examined, there are pregnancies which pose a serious problem for the mother; and, indeed, in those cases, it is necessary to consider all that can be done for both mother and child. As regards the more general claim that the tragic abuse of "back street abortions" awaits, like a predator, if there is an increase in the protection of the child in the womb, frozen or otherwise designated for experimental exploitation, the answer lies in providing the help that everyone needs, at whatever time, that will benefit all concerned.

As regards change, we do not know the future but we do know that the world is not a settled place and that, as we have seen, both old therapies are being reconsidered and new ones developed; and, therefore, the plight of the

---

[253] Cf. "The European Parliament calls for the inclusion of abortion in the Charter of Fundamental Rights of the European Union", by Bioethics Observatory | August 29th, 2022: https://bioethicsobservatory.org/2022/08/the-european-parliament-calls-for-the-inclusion-of-abortion-in-the-charter-of-fundamental-rights-of-the-european-union/44189/.

child conceived ectopically is not without a new context that challenges the almost total lack of interest in the life of the child. Thus a return to a more evidence and care based medicinal mentality will, hopefully, really motivate the provision of realistic help for both mother and child.

More widely, politicians, heads of organizations and activists come and go – not just because of retirement and natural death but also because of a word which has changed their life and opened them up to the possibility of a fullness of life beyond our own plans and projects (cf. Jn 10: 10)[254]. Truth, however, cannot contradict truth[255]; and, if we follow its gentle lead[256], we will be freed to live – both fully here and with the horizon of the

---

[254] Cf. Cerith Gardiner, 25/08/2022, "Breaking news: Actor Shia LaBeouf converts to Catholicism": https://aleteia.org/2022/08/25/breaking-news-actor-shia-labeouf-converts-to-catholicism/.

[255] "Truth Cannot Contradict Truth": Address of Pope John Paul II to the Pontifical Academy of Sciences (October 22, 1996): https://www.newadvent.org/library/docs_jp02tc.htm.

[256] Cf. *Dignitatis Humanae*: https://www.vatican.va/archive/hist_councils/ii_vatican_council/documents/vat-ii_decl_19651207_dignitatis-humanae_en.html.

heaven to follow. Who knows, then, what change will bring – even in the United Kingdom?[257] But unless change is consistent, like the grain of wood through which pass the forces of the universe, then change will not open upon new life but contradict it; and, therefore, it is essential to base our grasp of reality on the truth that ever opens upon a new future, forgiving and going forward!

Change, however, can start from the slightest positive beginnings and, as such, a fine example is a statement from judgement of the European Court:

The European Court 'argued in its justification that while the notion of property could not be limited merely to material objects, human embryos could equally not be reduced to the level of an object'[258].

---

[257] "Therese Coffey's views on abortion concerning, charity says", by Ione Wells, 07/09/2022: https://www.bbc.co.uk/news/uk-politics-62805268.

[258] *Ibid*, p. 131: European Court of Human Rights, Case of Parrillov. Italy, Judgment, August 27, 2015, http://hudoc.echr.coe.int/eng?i=001-157263 (accessed October 15, 2019) – as per the original author's document.

Thus we have the precedent, as it were, of a philosophical discussion which could go to the roots of a new declaration of human rights, beginning with the recognition that 'human embryos could equally not be reduced to the level of an object'. What, then, is a human embryo if it is not to be 'reduced to the level of an object'?

Part III:
The Disruptive Power of the Word of God

St. Thomas Aquinas says that an erroneous conscience is one which a person *believes* 'to be a right conscience' when it is wrong[259].

When, then, we are mistaken in our judgement and, therefore, consider an abortion a good act and, unable to be convinced by the evidence, there is the complementarity of faith and reason as expressed in the power of the word of God. Three successive popes have adverted to this power of the word of God to help us.

St. John Paul II says, in *Veritatis Splendor*, 'before feeling easily justified in the name of our conscience, we should reflect on the words of the Psalm: "Who can discern his errors? Clear me from hidden faults" (*Ps* 19:12). There are faults which we fail to see but which

---

[259] St. Thomas Aquinas, *Quaestiones disputatae de veritate*, q. 17, a. 4, translated by Fr. George Woodall and as supplied by him as part of the course content of the Maryvale Institute's now discontinued MA in Catholic Theology (2 pages).

nevertheless remain faults, because we have refused to walk towards the light (cf. *Jn* 9:39-41)'[260].

Cardinal Ratzinger, now Pope Emeritus Benedict XVI, citing the tragic blindness of those who have caused the death of many, goes on to say that this 'should rouse us to take seriously the earnestness of the plea: "Free me from my unknown guilt" (Ps 19:13)'[261].

Pope Francis, too, speaks of 'being "unsettled by the living and effective word of the risen Lord"[262]; and, therefore, he is speaking of the 'word of the risen Lord' as a unique word which can pass us to where we cannot take ourselves without His help.

Let us not underestimate, then, the proclamation of the word of God and its helpful, disruptive effect, guiding us to the full truth from whatever may be our starting

---

[260] *Veritatis Splendor*, 63.

[261] Cardinal Ratzinger, "Conscience and Truth", *Communio*, 37, (Fall, 2010), p. 538.

[262] Quoted from *Gaudete et Exsultate*, 137, by Pope Francis in his "Letter to Priests", https://www.hprweb.com/2019/09/letter-of-his-holiness-pope-francis-to-priests/#fnref-24676-26.

point[263]; indeed, if the Lord is with us who can be against us (Rm 8: 31)? At the same time, let us remember that the Lord wants no one to be lost (cf. 2 Peter 3: 9); and, therefore, whether it is the child, the mother, the father, the scientist, the administrator, the lawyer, the engineer – the Lord seeks the salvation of us all!

The Human Unpredictability of an Answer to Prayer: The Will of God

We live, then, in a relationship to God – to God who calls us constantly and, when it comes to the end of our life, He will say: "What more could I have done for you?" (cf. Is 5: 4); and, therefore, let everyone examine their life and ask, searchingly, have I heard and denied the voice of the Lord? Many times He called to me, from an inexplicable love of the word of God, the Scripture, which I never opened apart from the times I read from it as a child in the Church; but then, too, when I was in the process of

---

[263] This is altogether more explored in Chapter Seven *of Human Nature: Moral Norm*: forthcoming from En Route Books and Media in 2022: https://enroutebooksandmedia.com/humannature/.

committing suicide, swallowing pills, at 14, I "saw" the possibility of coming into the presence of Christ and the apostles and I was afraid and started drinking water. This "coming of Christ" into my life went on until, at 40, I understood and believed that "If God can create everything out of nothing He can make a new beginning for the sinner" (CCC, 298) – and that sinner was me!" Now, at 66, having married at 40 and 25 years on, with 8 children, three in heaven, one there because of an abortion years before my married life, it is clear that God loves the sinner and wants no one to be lost (cf. 2 Peter 3: 9)[264]. As for myself, the need of God does not diminish but becomes clearer, as I have to go into hospital for a sixth time, for a fourth operation; and, as you can imagine, marriage and eight growing children give us plenty to pray about, both widening our prayers to include all growing up and what is going on in the world!

---

[264] Cf. The Quartet: *The Prayerful Kiss: A Collection of Poetry and Prose* (I); *Honest Rust and Gold: A Second Collect of Prose and Poetry* (II); *Within Reach of You: A Book of Prose and Prayers* (III); and *The Family on Pilgrimage: God Leads Through Dead Ends* (IV). All the books are on Amazon.

Chapter Ten: A Conclusion in Three Parts 323

There are, more widely, numerous testimonies of those who have been helped by God in the challenges of life, some of which are truly extraordinary; and, therefore, let us avail ourselves of both the help of God and help of the testimony of those who have experienced it[265].

Let us avail ourselves of new initiatives, whether it be a pro-life hospital[266] or one waiting to be discovered in the heart of God, and implore the help of the Holy Spirit in developing them! And let us not forget the immense fold of those torn from life who, in their simplicity and love,

---

[265] There are many testimonies but, for brevity's sake, these two will suffice: "Pregnant Mom Diagnosed With Cancer Rejects Abortion, After Birth Now She's Cancer-Free", National, *Micaiah Bilger*, Sep 6, 2022: https://www.lifenews.com/2022/09/06/pregnant-mom-diagnosed-with-cancer-rejects-abortion-after-birth-now-shes-cancer-free-2/; and, secondly, there is the truly challenging experience of a young couple, called *Chiara Corbella Petrillo: A Witness to Joy*, translation by Charlotte J. Fasi, Manchester, New Hampshire: Sophia Institute Press, 2015.

[266] John Burger, "A new Catholic medical school seeks to restore a culture of life", 22 September, 2022: https://aleteia.org/2022/09/22/a-new-catholic-medical-school-seeks-to-restore-a-culture-of-life/.

appeal for all – both those still tearing up the defenceless and those who are striving, day in and day out, to speak and act for a love that forgives and builds anew the culture of life.

Let us avail ourselves of the power of prayer[267].

---

[267] Cf. In the UK, "Doctor vindicated as NHS England allows him to pray with patients", 26th September, 2022: https://christianconcern.com/news/doctor-vindicated-as-nhs-england-allows-him-to-pray-with-patients/?utm_source=Christian+Concern&utm_campaign=3cde71d748-BN-20220928-Scott&utm_medium=email&utm_term=0_9e164371ca-3cde71d748-127772892.

A Very Distinguished Testimony
Saint Teresa of Calcutta

"I was surprised in the West to see so many young boys and girls given to drugs. And I tried to find out why. Why is it like that, when those in the West have so many more things than those in the East? And the answer was: 'Because there is no one in the family to receive them.'

"We are talking of love of the child, which is where love and peace must begin. These are the things that break peace.

"But I feel that the greatest destroyer of peace today is abortion, because it is a war against the child, a direct killing of the innocent child, murder by the mother herself. And if we accept that a mother can kill even her own child, how can we tell other people not to kill one another? How do we persuade a woman not to have an abortion? As always, we must persuade her with love and we remind ourselves that love means to be willing to give until it hurts. Jesus gave even His life to love us.

So, the mother who is thinking of abortion, should be helped to love, that is, to give until it hurts her plans, or her free time, to respect the life of her child. The father of that child, whoever he is, must also give until it hurts.

By abortion, the mother does not learn to love, but kills even her own child to solve her problems. And, by abortion, that father is told that he does not have to take any responsibility at all for the child he has brought into the world. The father is likely to put other women into the same trouble. So abortion just leads to more abortion. Any country that accepts abortion is not teaching its people to love, but to use any violence to get what they want. This is why the greatest destroyer of love and peace is abortion.

Many people are very, very concerned with the children of India, with the children of Africa where quite a few die of hunger, and so on. Many people are also concerned about all the violence in this great country of the United States. These concerns are very good. But often these same people are not concerned with the millions who are being killed by the deliberate decision of their own mothers. And this is what is the greatest destroyer of peace today — abortion which brings people to such blindness. And for this I appeal in India and I appeal everywhere — "Let us bring the child back."

"Please don't kill the child. I want the child. Please give me the child. I am willing to accept any child who would be aborted and to give that child to a married couple who will love the child and be loved by the child. From our children's home in Calcutta alone, we have saved over 3000 children from abortion. These children have brought such love and joy to their adopting parents and have grown up so full of love and joy." (This appeal, then, is made in the name of all who are open to the possibility of welcoming life![268]).

---

[268] In whatever country a person is in there will be, generally, organizations willing to help, whether religious or otherwise e.g. in the United Kingdom, SPUC (https://www.spuc.org.uk/About-Us/About-Us/Contact-Us) ;Right to Life (https://righttolife.org.uk/contact-us); 40 Days for Life (https://www.40daysforlife.com/vigil-search.aspx); The Good Counsel Network (https://www.goodcounselnet.co.uk/40-Days-for-Life); Life (https://lifecharity.org.uk/) or contact your local parish church and ask etc.

"If we remember that God loves us, and that we can love others as He loves us, then America can become a sign of peace for the world. From here, a sign of care for the weakest of the weak — the unborn child — must go out to the world. If you become a burning light of justice and peace in the world, then really you will be true to what the founders of this country stood for. God bless you!"[269]

---

[269] "Blessed Mother Teresa on Abortion": "At the National Prayer Breakfast in Washington, D.C, on February 5, 1994, Mother Theresa dared to speak her mind and heart about the right to life": https://www.catholicnewsagency.com/resource/55399/blessed-mother-teresa-on-abortion.

WHERE TO GET HELP?
"SEEK AND YOU WILL FIND" (MT 7: 7)

HELP WITH PREGNANCY OR POST-ABORTION COUNSELLING

Dr. Pat Castle: '*78% of post-abortion mothers said if they had encountered ONE supportive person or encouraging message, they would have chosen life.*' And, therefore, he founded "Life Runners": they wear '"REMEMBER The Unborn" jerseys as a public witness in over 3,300 cities'[270].

Whatever country a person is in there will be, generally, organizations willing to help. Whether you google a parish or a diocesan website, life, pregnancy help, pro-life organizations, you will find they exist almost everywhere.

---

[270] ITEST "A Post-Roe World" Webinar publicity, now with the WCAT TV presentation on it: https://faithscience.org/post-roe/.

If you are afraid to ask for help pray, open a Gospel at random, and then ask for help.

The Sister of Life, America, go to "Where We Are"; they are in six states and Toronto, Canada: https://sistersoflife.org/

The Sister of Life, Glasgow, go to the contact page and also look at the Links, further down: https://gospeloflifesisters.wordpress.com/pregnant-we-can-help-you/

SPUC: https://www.spuc.org.uk/About-Us/About-Us/Contact-Us

See also "The Alma Mater Fund" for pregnant university students across the United Kingdom: https://www.spuc.org.uk/Article/385354/The-Alma-Mater-Fund-expands-across-the-UK.

Right to Life: https://righttolife.org.uk/contact-us

40 Days for Life: https://www.40daysforlife.com/vigil-search.aspx

The Good Counsel Network: https://www.goodcounselnet.co.uk/40-Days-for-Life

Life: https://lifecharity.org.uk/

Or contact your local parish church and ask

A Testimony from a Man for Men

Indelible (I) and the Rise of Articles on St. Joseph and Fatherhood (II)

In the testimony that follows, truth and love belong together and, being together, are completed by forgiveness and hope. If, then, you need help, contact one of the organizations on the previous pages.

Indelible (I)

This poem is called "Indelible" and is about my loss of a child to abortion and is taken from the first collection of prose and poetry called *The Prayerful Kiss (A Collection of Poetry and Prose)*[271].

---

[271] Francis Etheredge, *The Prayerful Kiss*: https://enroutebooksandmedia.com/theprayerfulkiss/; and what follows is an excerpt from the prose introduction to the poem and, therefore, there is additional testimony in the book on pages 49-54.

An Unexpected Joy: An Unprecedented Pain

Sin does not describe the experience of suddenly, unexpectedly, discovering an inexplicable joy that arises out of the conception of a child; indeed, however aware we were of the possibility of conceiving a child, there came an unforgettable joy from the very roots of human being. Whatever the "noisy" claims about a "clump of cells" - the reality of parenthood had an unmistakable beginning: a trumpeting joy. It is possible to understand human psychology as if our whole being is a kind of self-originating expression of conscious reactions; however, given that the very existence of each one of us arises out of relationships, it is not possible to understand ourselves except "through" relationship. Perhaps we need to recognise, then, that seeing is seeing something or someone: that there is a kind of interior dialogue between ourselves and what exists. Consciousness is not just about "admitting" the presence of a self – it is also about dialoguing with what is real. Thus, on reflection, the joy that arose was as indistinguishable from the very coming into existence of "another" as it was unbidden; indeed, as surrounded as this moment was by all kinds of difficulties and uncertainties, it is extraordinary that it was joy that rang out.

A Testimony from a Man for Men

Joy and pain express "relationship"

But then the child was aborted.

INDELIBLE

Joy announced you to me.

I know the smothering imperfection of our time:
the anxious lying of single people –
am I a father? am I a mother?
do we have a child?

"Your" existence your mother denied:
her lie to my uncertainty, a slipperiness supplied –

and my heart filled,
as I foresaw her,
beside me,
stretching out our tent:
swelling,
growing rounder
and my hands and body knowing
life between us growing

from within us both beginning
but now in her becoming -

And again I pulled apart:

"Are we married?
Have we the right?
To love as lovers and unite?"
But then I saw, as in a sleepless dream,
in an in-hospitable place, being done what we cannot undo:

separating what God had joined together.
Your mother and I met again,
and her secret, she shared:

That she thought you were a "blobby mass";
and, because of it,

You are no longer where you were.

Listening was a splashing pain,
a splintering, swiftly slicing pain.

"I'm sorry" was a word too heavily burdened to be spoken. We named and prayed for you.

Too personal, I know, to make public, but public it now is, because too many died, before you,
and you did to me what millions did not do.

I was wrong, I know, to love when love was not committed, between your mother and me.

Listening was a splashing pain,
a splintering, swiftly lashing pain.

And I pulled apart again - hurting hard again:

a hidden shard of pain,
seizing leaving as a means
to hurt her hard again.

.

"I am sorry": a word that grew in me to speak.

Listening was a splashing pain,
a splintering, swiftly swiping pain.
"I am sorry" is a word to strengthen the weak.

Reconciled
Your mother and I parted.

The dead Christ-child lay in the arms of Mary:
a blooming brightening
unforgettable being.

And out of Love's many-chambered-petalled-heart,
our child spoke like scent:

"Look and see the Crucified:
His Resurrection is our new life.
Do not escape your suffering and
it will give you life again.

When Easter crowns the crucifixion and
completes the gift of Christmas –
we will go up in song,
and love will cry out:
Amen in song!
Amen in sight!
Amen!

Articles on St. Joseph and Fatherhood (II)

There are no doubt many articles on St. Joseph and Fatherhood on the internet; however, for brevity's sake, I include a few of my own and the website of "The Catholic Weekly", Sydney, on which it can be found, particularly because this website has a number of others on the theme of fatherhood and St. Joseph. In a word, the work of fatherhood is a work of ongoing conversion and, with all that has been said in this book, there is help available.

Francis Etheredge:

"Lord: Do You Mean Me? A Father-Catechist!": https://www.catholicweekly.com.au/lord-do-you-mean-me-a-father-catechist/

"Fatherhood: Failure or Promise?": https://www.catholicweekly.com.au/fatherhood-failure-or-promise/

"Parents as First Teachers of the Faith": https://www.catholicweekly.com.au/parents-as-first-teachers-of-the-faith/

"St. Joseph: Model Father": https://www.catholicweekly.com.au/st-joseph-model-father/

On this website there are a host of other articles on St. Joseph and fatherhood – use either of them as a search term.

Further Reading

A Variety of Prior Work on Conception

The argument advanced in this work is also explored in numerous articles and books, all of which entail their respective sources.

The two articles below were published by the Catholic Medical Quarterly, UK, and were also included in the book, **Scripture: A Unique Word**: https://www.cambridgescholars.com/product/978-1-4438-6044-4:

Part I of II: "A Person from the first instant of Fertilization", *Catholic Medical Quarterly* (August 2010, Vol. 60, No. 3, pp. 12-26)

Part II of II: "A Person from the first instant of Fertilization", *Catholic Medical Quarterly*, (November 2010, Vol. 60, No. 4, pp. 20-26)

Chapter 7 of **The Human Person: A Bioethical Word**: https://enroutebooksandmedia.com/bioethicalword/.

Chapter 5, Part II, of ***Conception: An Icon of the Beginning***, was written by two specialists: Professor Justo Aznar (a former member of the Pontifical Academy for Life) and Julio Tudela; this Chapter gives excellent evidence of the earliest interactions between the sperm and the ovum. In other words, no sooner do sperm and egg come into contact with one another than they start to interact and from then on everything is relevant to the formation of the human embryo: https://enroutebooksandmedia.com/conception/

As regards the theological argument, go to Chapter 5 of ***Mary and Bioethics: An Exploration***, this gives the reasoning that led to the view that Mary cannot be wholly holy unless there is an instantaneous union of the body and soul from the first instant that each exists and exists as one: https://enroutebooksandmedia.com/maryandbioethics/.

But see also, ***Reaching for the Resurrection: A Pastoral Bioethics***, which, in Chapter Six, considers our beginning and end in the light of Christ's conception and death: "Brain Death and the Life and Death of Christ: Can the Living Be Dead?":

Further Reading

https://enroutebooksandmedia.com/reachingfor-theresurrection/

Articles published by the *National Catholic Bioethical Quarterly*:

"The Mysterious Instant of Conception", *National Catholic Bioethics Quarterly* of America, Vol. 12, Autumn 2012, No. 3, pp. 421-430.

"Frozen and Untouchable: A Double Injustice to the Embryo", *National Catholic Bioethics Quarterly* 16.1 (Spring 2016).

"The First Instant of Mary's Ensoulment", *National Catholic Bioethics Quarterly*, Vol. 19, Autumn 2019, pp. 359-367.

Article published in "Ethics and Medics":

"On Regulating IVF" in "Ethics and Medics", Volume 41, Issue 7, July 2016

In general, I would like to thank all those who have collaborated over the years and, in many cases, continue to do so; for, in the end, while this work is my own, the extent to which it draws on the work of others is almost, indirectly, incalculable.

And Specific Documents of the Catholic Church in English, albeit with a Latin title taken from the first words of the Latin text

> *Humanae Vitae*: https://www.vatican.va/content/paul-vi/en/encyclicals/documents/hf_p-vi_enc_25071968_humanae-vitae.html.

> *Donum Vitae*: https://www.vatican.va/roman_curia/congregations/cfaith/documents/rc_con_cfaith_doc_19870222_respect-for-human-life_en.html.

> *Evangelium Vitae*: https://www.vatican.va/content/john-paul-ii/en/encyclicals/documents/hf_jp-ii_enc_25031995_evangelium-vitae.html.

Dignitas Personae: https://www.vatican.va/roman_curia/congregations/cfaith/documents/rc_con_cfaith_doc_ 20081208_dignitas-personae_en.html.

Catechism of the Catholic Church, particularly paragraphs 2270, 2274-2275: https://www.vatican.va/archive/ ENG0015/_INDEX.HTM.

www.ingramcontent.com/pod-product-compliance
Lightning Source LLC
Chambersburg PA
CBHW071109160426
43196CB00013B/2511